FUTURE REVOLUTIONS

FUTURE REVOLUTIONS

Unravelling the Uncertainties of Life & Work in the 21st Century

DAVID MERCER

This edition first published in Great Britain in 1999 by
Orion Business
An imprint of The Orion Publishing Group Ltd
Orion House, 5 Upper St Martin's Lane, London WC2H 9EA

A CIP catalogue record for this book is available from the British Library.

ISBN 0–75281–378–1

Printed and bound in Great Britain by
Butler & Tanner Ltd, Frome and London

CONTENTS

1

REVOLUTIONARY TIMES

As we look around us, at the end of the second Millennium, it might seem as if the end of the world is at hand. All – in the popular media at least – seems doom and gloom. Confusion – if not outright anarchy – rules. Fintan O'Toole – a columnist with the *Irish Times* – contrasts it with the widespread optimism at the end of the nineteenth century. Then, 'the future was held by the likes of Sherlock Holmes [and] the application of scientific methods to the problems of the present would allow for a benign outcome in the future.' The black pessimism of our own times, on the other hand, has rather different outcomes: 'Our equivalents of the sleuth of Baker Street are Judge Dredd, Robocop, Blade Runner and Terminator, mythic figures in whom futuristic technology and mediaeval visions of hell have come together to form a nightmarish anti-utopia.'

Indeed, we have had mass genocide in Rwanda, intractable civil war in Bosnia, genuine anarchy in Somalia, wholesale unemployment around the world, and the underclasses setting up independent – drug-funded – rule in the inner cities of the United States. The family is breaking up; and more than three hundred organizations, in the United States alone, look forward to seeing Armageddon arrive at the end of the Millennium. Even Nostradamus forecast five hundred years ago:

In the year 1999 and seven months,
from the sky will come the great king of terror.
He will bring back to life the great King of the Mongols.
Before and after, war reigns happily.

Henry Porter, writing in *The Guardian*, colourfully sums up the position as 'all the forecasts and projections are pervaded with the cheeriness of a Hieronymus Bosch painting.'

There is, indeed, some substance to the popular view that the very foundations of our world are being shaken. As we will see, the information technology (IT) communications revolution means that our working lives will never be the same again; the economic effects of globalization are being experienced everywhere; social revolutions are rocking whole communities, and destroying the family; the demise of the nation state, and of its political parties, really is posing a threat to all politicians. It is no wonder that most of our leaders are just as confused as we are, and have no sensible answers to the problems we face and increasingly retreat into comforting nostalgia about past values, while adopting corrupt practices to try – without success – to save their own skins. *What hope is there for the future of humanity?*

In fact, as this book reveals, the future – even in the relatively short term – should be viewed not with the *pessimism* we are seeing in the media but with *optimism.* Thus, we already have a genuine global peace, with the nuclear threat largely removed, for the first time in half a century; even Southern Africa, with the blight of apartheid removed, is peaceful; the Third World in Asia – including, most notably India and China – is rapidly advancing towards developed status; dictatorships everywhere are being undermined by the new international harmony, and by the 'truths' being transmitted via satellite television. People in almost all countries are earning more – in the West the IT revolution is making most of us significantly richer – and living longer; and the individual rights of mankind, and especially of womankind, are being recognized and protected. Thus, for the majority of us, life has never been better!

That the future will be even better is not just my own personal view but the overwhelming, considered view of the representatives of more than a thousand organizations who have worked with us over the past five years on the research which underlies this book. We have gathered the expectations of what the future of humankind might be from managers in most of the world's largest multinationals as well as from government ministers and leaders of international bodies across the world. These have been well considered decisions, the subject of intensive sessions lasting between three and thirty hours each, representing the expectations of the large organizations which will decide the future of the world over the next decades.

Thus, the justifications for the views I put forward are not my own, but are those of the many hundreds of individuals who each gave so much time and energy to the work; and for that I am enormously grateful to them. For the reader it is also the guarantee that what this book has to say has the authority necessary for you to confidently use the map of the future it describes as the basis for your own planning. It is the authority of those hundreds of contributors which should assure you that the views reported in the book are genuinely meaningful. You should refer to the long list of acknowledgements, at the end of this book, to see who the main influences have been. As such, the forces which are described in the book, and whose likely outcomes are mapped, are those which are already shaping the future. Indeed, since most of the individuals and groups involved clearly shared a common view of aspects of that future, the results almost certainly spell out what will happen in that future. If these thousand or so key organizations are optimistic about the future, then the prospects for that future, for all of us, look good; and don't let any doomsayers tell you otherwise!

Beyond our own research, the quoted contributions have proved to be important at two levels. First, they fill gaps in coverage, where our groups did not have the requisite expertise to comment. Second, the comments made by these other 'futurists' also show a remarkable degree of convergence with each other and with the comments made by our own groups. They have demonstrated that the overall picture of the future, seen from a variety of different perspectives, is the subject of widespread agreement.

We live in truly revolutionary times. They are the most exciting – if not the most confusing – that humanity has ever experienced. Christopher Farrell, in *Business Week* (12 December 1994) succinctly describes the position thus:

> On the eve of the 21st century, the signs of monumental change are all around us. Chinese capitalists. Russian entrepreneurs. Nelson Mandela President of South Africa. Inflation at 7% in Argentina. Internet connections expanding by 15% per month. Fibre Optics transmitting 40 billion bits of data a second. From government dictators to assembly line workers, everyone seems aware that unfamiliar and unusually powerful forces are at work. Says Shimon Peres, Israel's Foreign Minister: 'We are not entering a new century. We are entering a new era.' A great transformation in world history is creating a new economic, social and political order.

There may not be blood on the streets, but over the next few decades, life on Earth will experience more change than it has in several centuries past. For some, especially 'members of the establishment', these revolutionary changes will create anxiety; and the symptoms of such stress became very evident as the 'Crisis Decades', starting in the late 1970s, progressed. For the great majority, however, these changes will herald the coming of something approaching utopia. At present, though, that majority is understandably unnerved by the bewildering lack of leadership being shown by their politicians. One message of this book is, therefore, not just a forewarning of the shattering changes which are taking place but of the real hope they portend for most of humanity.

In the West, which will continue in the short term to lead the development of society, the main changes will affect the very fabric of that society. Unfortunately, in our present age of uncertainty we are as yet aware of only parts of these changes and do not see the complete picture. We interpret the parts as being the end of society as we know it, but they are in reality the start of something much more positive. Thus, work, leisure and even the hallowed institution of the family are not disintegrating but are all metamorphosing into radically new forms; above all, we are at long last gaining control of our own destiny. The new form of the community in which we are to be located may as yet be uncertain, but the pain we now feel, and which we find so distressing, is not a symptom of a terminal decline but the inevitable accompaniment to progress.

The power of the mature individual will be experienced in portfolios of lifestyles, allowing each of us to develop ourselves to the full, and in portfolios of issues. It is no wonder that politicians, and their establishment partners in the mass media, face the future with trepidation.

On the wider, global, scene, the changes will be even more dramatic. The citizens of the Third World are, in ever growing numbers, joining those who are already enjoying the fruits of development. For the first time, the majority of humanity will see a future which promises an increasing quality of life. In a world which now values individuals equally, in terms of their buying power almost as much as their votes, the shift in power to the masses will itself have dramatic consequences. At the other extreme, however, we will – as the most important collaborative venture of the new age – see the start of the long-delayed colonization of space; to give impetus to a, literally, ever-expanding future. Nobody may have gone before into these as yet inhospitable domains, but millions will soon do so; and billions will eventually find their home there.

In previous times, such thoughts might have been dismissed as, literally, utopian. But now, for the first time, we can afford to realize even our wildest dreams. The potential resources at our command already exceed anything we might need to create that utopian future. The well publicized fears of the environmentalists are exacerbated by what they see as the inevitably bad behaviour of humanity. This book – covering a much wider range of issues than what they focus on, and with a much better appreciation of what expectations others hold – makes the (fully justified) assumption that the long-term future of humanity will, however, be created by its good behaviour. We will positively choose to make the correct decisions about the environment. Indeed, the widely reported fears of environmental catastrophe are, while highly exaggerated, a very visible reassurance that we *will* resolve the problems facing us. Our research results show that they are a positive sign of our determination to take the necessary actions.

In this context, the challenge is now to redistribute the resources to where they are needed, and then to manage them so that their use might be optimized and not, as at present, squandered. We already have the food, energy and mineral resources which are needed for the immediate future, and those needed for the longer term await development in space if not on Earth. Above all, however, we have now created, even in most parts of the Third World, the ultimate resource needed: an educated population. Given a humankind which does choose to make the correct political decisions, the future is unlimited.

The outcome is that social forces, almost by themselves, will now shape that future. What people want, and – perhaps more important – what they expect, to happen will largely determine what will happen! By capturing people's expectations of the future we can now predict that future. Where the considered expectations of more than a thousand representatives of the major organizations, which already run our world, have been gathered – as they have been for this book – we can indeed begin to have some confidence in such predictions, even in terms of some of the details.

An even more important message is, however, that the future will now be what we choose to make it! James Ogilvy (1992), one of the founders of the Global Business Network, suggests that 'Simply to be a human being is to be a futurist of sorts, for human freedom is largely a matter of imagining alternative futures and then choosing between them'.

Revolutionary Trends

I rather understated the drama about to unfold when I referred to it as revolutionary. There are, in fact, a number of different revolutions taking place which are each due to reach their separate climaxes in the coming decades. Each of these might be argued to be as important as the previous revolutions which have unsettled humanity. Coming together, and being leveraged by the powerful psychological impact of the Millennium, it is not unreasonable that there will be an accompanying, dramatic series of changes in society.

The first of these revolutions is that which has been most widely described as the IT Revolution, though it might now more accurately be titled the Communications Revolution since it is the wider impacts brought about by the dramatically enhanced powers of individual communication which will ultimately have the greatest effect on society. This is part of a general revolution in technology, which means that we have to hand, albeit not always yet in production, the technology necessary for humanity's development over the next half-century. It is very difficult to conceive of any programme of development which might be held back because the technology was not available.

The second set of revolutions are in the field of sociology. Although they tend to parallel each other, they can be most meaningfully seen as three separate revolutions. The first of these carries the often quoted title of 'Post-Modernism'. 'Modernism' is now quite well established as a phase in development, the form of society which emerged after the first Industrial Revolution, and which has provided the context for most people's lives until recently. In the historical context, Post-Modernism simply is the next stage of development beyond this. I prefer to characterize it above all as the empowerment of the individual.

The second social revolution, which has been promised by some for a number of decades, is that of 'post-materialism'. This takes as its starting point the fact that many individuals have now reached satiation in terms of the goods they own: the usefulness of the third or fourth car must surely be a lot less than that of the first. The expected result is that the focus of their purchases will shift to the non-material, towards a more inwardly directed, not to say more spiritual, life. Whether these hopes of a 'better' society will be realized is questionable, but what cannot be challenged is that society is certainly moving away from the acquisition of physical goods as the prime source of demand. Not least, this is reflected in the shift in patterns of employment to meet the rapidly increasing demand from the service sector.

The third social revolution relates to the patterns of work. It is sometimes referred to as 'post-Fordism'. The symbol of the modernist period was the production line, where dehumanized workers were driven by, and almost became part of, the machine. Now, the move to the information society, and the parallel move to service work, means that the individual – using his or her intelligence flexibly – has become the prime generator of added value. Thus, the individual has become the most important investment, especially in terms of the education and training they have received. This has long been recognized by the Japanese corporations, and is now being emulated by the more advanced organizations in the West.

Resources

As has already been stated, the key to these developments will be the fact that, for the first time, the resources are now available to achieve almost all that society might want. Those resources are, though, unevenly distributed and government intervention, international and supranational, will be needed to rectify this. Subject to this redistribution, in general there is already enough food to feed everyone, enough mineral resources to supply industry and enough energy to power it – albeit not at the sort of artificially low prices which led to the wasteful excesses before the 1970s. In any case, in outer space there is all of this in abundance, not least almost free solar power.

Above all, there is now sufficient educated manpower to take advantage of these physical resources. It is humanity itself, especially those members with special skills, which now is in the shortest supply. On the other hand, many countries, including some in the Third World as well as those in the Developed World, have invested heavily over recent decades to educate their populations to the level needed to create the unlimited future this book describes.

An especially important effect, in the West at least, is that the emancipation of women is now going beyond mere equality, even to establish a form of supremacy, so that the beginning of the Third Millennium might be considered to be the age of women. Not merely are the 'feminine' societal values coming to the fore, but the new female stereotype, with its emphasis on education, better fits them for the intellectual demands of the new information society.

Indeed, the IT Revolution is becoming a major driver for change. It is not, though, the technology itself which matters most. Despite all the hype, the appearance of the PC on so many desks has not made a major impact. So far it has been typically used as a replacement for the typewriter and calculator and has been most often acquired as a status symbol. The major impact is only beginning now that these millions of PCs are being connected to each other. When they connect directly to our brains, as some scientists are predicting, the power of human beings, not just computers, will grow even more dramatically, creating a new stage of evolution – *homo integrans*!

This form of communication is genuinely revolutionary. It expands the horizons of the individual by an order of magnitude. Much has been made of the ability to 'talk' to people on the other side of the globe, but the real benefit will initially come from the ability to talk simultaneously, and efficiently, to tens of people in the same office as yourself. Even without the direct interface with our brains which is predicted, it still represents a genuinely new form of communication which will enable many more people to gain access to the discussions which have previously taken place between a few privileged individuals. Much of the language of this new medium remains to be developed, so its power has not yet been really tested; but the potential for widening both the 'senses' of the individual and their knowledge, especially that held personally by others rather than impersonally in libraries, can already be seen to have truly revolutionary potential.

The impact upon society as a whole will be no less great. In organizational terms, the most important new communications flows will be horizontal rather than vertical. This has already started to undermine the traditional role of management, but it will go farther, to demand management structures which are very close to those of the Japanese corporations. This will result in the consequent reinforcement of the view that individuals working in an organization be recognized as its most important investment.

At the personal level, it will massively reinforce the ability of the individual to build his or her own tailored portfolio of lifestyles. This is where the expansion of horizons may be most adventurous – especially when the new super-highways allow the easy, and cheap, transmission of video material as well as text. The new groups, to which they can belong, may no longer be restricted to members gathered from the surrounding communities; they really will be able to contain members from around the world. This is already how physicists researching high-energy particles operate – they even run their experiments from thousands of miles away – but in future it may be just as easily how champion growers of leeks communicate with fellow aficionados!

A consistent aspect of the scenarios emerging in my investigations has been a persistent worry about 'disasters'. These typically see humanity wiped out by an uncontrollable event, now more typically a natural disaster rather than a man-made one – a comet hitting the Earth or a new, unstoppable disease rather than global warming or a nuclear war. This is an especially catastrophic version of the range of environmental disasters which have more generally dominated the headlines in the popular press. According to our research, however, the favourite insurance against such threats is now to spread humanity *beyond* the bounds of Earth, most memorably described by the writer of one of our group scenarios as 'God selects a second Ark – a Space Ark'. The real statistics behind such risks may not justify the development of space, but the powerful – albeit somewhat irrational – fears of the public still represent a powerful argument for colonization of other worlds. At the same time, while humanity is moving towards a global society and, ultimately, global political union, space may represent a very important unifying focus for this thrust.

Global Individualism

The emergence of the individual is rapidly becoming a worldwide phenomenon. Empowerment of that individual, especially of women, has become a touchstone of the development of a country's civilization. Confusingly, it is often described in terms of empowerment of just the individuals in the élite, the friends of the establishment. These are to be freed of constrictions, to make almost obscene profits at the expense of others. Rabid capitalism of this type has been the byword of many governments in recent decades. On the other hand, the same messages are, unnoticed by many of the politicians promoting them, just as avidly being absorbed by the many other individuals who would previously have been consigned to a mere role as part of the 'masses'. These new voters are starting to demand the genuinely individual rights which have been, albeit rhetorically, already offered.

The new Right would claim that this growing individualism has removed the need for public provision; the market should, and will, provide all – privatization is the only solution. Margaret Thatcher memorably summarized this view when she claimed that there was 'no such thing as society!' The result is that, during the crisis decades at the end of the twentieth century, much of the role of the community has been dismantled. Some would go so far as to say that the social infrastructure has been more gratuitously vandalized by disoriented governments themselves than by the underclasses which they have created, and which they so fear.

The paradox is that the new individual is in greater need of the community. Left alone, with only the market to mediate his or her aspirations, the result would be anarchy. Chaos would ensue as the billions of individuals demanded their own rights, regardless of those of others. This is, despite the free-market rhetoric of politicians, already recognized by the ever-growing number of regulations which seek to constrain the activities of all parts of society. Our research shows that even the management establishment, which in a free-market economy should be leading future developments, looks above all to government regulation to determine its future.

The reality is that a new relationship, a compact if not a contract, needs to be developed between the community and the individuals which it contains. On the other hand, the model of previous generations – whereby the community arbitrarily imposed the views of majority groups on everyone – no longer meets the needs of the individuals involved. What the new relationship will be is not yet clear to see, though the evidence is that it is already emerging, around the fringes of existing society, and it definitely will not be one prescribed by ideological extremes. It will no longer be sufficient dogmatically to provide just one solution to any problem. With the growing complexity of the needs of individuals, there will often need to be almost as many solutions as there are individuals involved. Before the information society this would have been impossible, but the new technology now offers the possibility of a viable outcome. Each of us can now, at least in theory – and soon in practice – have a unique relationship with the community to which we belong.

To a certain extent, however, it is no longer possible to describe just one community to which an individual belongs. In most Western countries the days when a geographic neighbourhood, and a related class or group membership, determined the community to which any individual belonged are long since gone. Indeed, neighbours are now one of the groups with which individuals see themselves as having little in common. Even the community of fellow workers is diminishing in importance, as the workplace loses its place as the focus of the individual's life. Instead, we now typically give a shifting allegiance in turn to each of the multiple groups to which we belong.

Depending upon the circumstances, we may indeed see our interests shared with very different, even seemingly contradictory, groups. At work our loyalty may still be to our fellow workers, but it might also be to the profession to which we belong or to the project group with which we are currently associated. As a consumer we might pose as a figure of fashion, or see ourselves as a protector of the green environment. In our free time we might be a member of a sports club, or a study group, or a local pressure group.

Indeed, the growing political importance of single-issue pressure groups most graphically illustrates how complex the process may become. In future we may see no contradiction in supporting a left-wing environmental group while, at the same time, we are an active member of a right-wing local conservationist group. This may offend the main political parties which have come to depend upon individuals committing their blanket support across all issues. It is, however, typical of the new portfolio approach being adopted by us; we now decide for ourselves, on each separate issue, what is the correct decision and refuse to sub-contract it to the establishment. We are already demonstrating our resolve by adopting new forms of family relationship; and it is these changes, unrecognized and unblessed by the establishment, which are significant.

In part, this new portfolio approach will be provided by the market as lifestyle marketing has already, in very crude form, attempted to do. The resulting many markets, and segments within them, will place new demands upon management. The rapid growth of computer-driven database marketing, which aims to simulate a unique, personal relationship between a supplier and each customer, is just one important example.

In the main, however, it will require an increased provision of public, community services. Many of our needs – not least our demand to live in a suitable environment – can only be met by the community. The market simply is not equipped to provide many of these services. The immediate need will be even greater, where these services are precisely those which were vandalized by governments during the crisis decades; though the availability of large numbers of unemployed, unskilled, workers – the category needed for such support – should aid the recovery.

As was mentioned earlier, perhaps the most potent driver will be the rapidly growing power of women. They, even now, are shaking off the constraints imposed by men and – as their traditional talents, let alone the new skills they are now learning, are already better suited to the demands of the new information society – they are starting to achieve the genuine equality of opportunity which should have been theirs long ago. More importantly, the whole of the new society – men as well as women – is beginning to accept increasingly 'feminine' values – co-operation rather than competition – so that, as claimed earlier, the twenty-first century may justifiably come to be called the women's century.

Global Power

Globalization is bringing a common viewpoint, and much the same set of shared values and aspirations, to peoples around the world. Television now reaches almost all communities; even the slums of Bombay are being wired up for satellite reception. Only the most remote Third World farming communities are as yet beyond its grasp. In this way, television is by far the most potent educative medium for the whole world. This carries its own risks, where the products of Hollywood are all pervasive. Its unreal lifestyles set the standards for impossibly happy family life in the West as much as elsewhere. This poses a degree of responsibility which the commercially driven entertainers ignore. The result is that the dream of many around the world is to live in downtown Los Angeles; a dream which those who have actually visited the place would more readily characterize as a nightmare!

In addition, as I mentioned earlier, redistribution of resources – around the world – should be one of the key features of the next century. Indeed, the new redistribution of political power will encourage a fairer distribution of economic power; and will persuade even governments and corporations in the West to divert rather more of their ever-growing resources to the development of the Third World. This will be for two reasons of considerable self-interest. Wealth increasingly depends upon market demand rather than supplier pressure, and – with more than 90 per cent of the world's population – the Third World will become the main economic driver in the Third Millennium. At the same time, the political power – driven by the same inexorable numbers of population – will also swing to the Third World. The same shifts of power will inevitably also lead to a change in the state of mind of the current superpowers, which no longer will be able to dominate global politics. As the pendulum starts to swing the other way, there is likely to be a scramble to put in place genuinely democratic supra-national institutions, ones which will protect the interests of minorities – as the Western powers will become – rather than continue to entrench the interests of the major players as does the United Nations. In any case, the power of the latter is gradually slipping away, as it follows the path of its predecessor – The League of Nations – into obscurity.

Supporters of world government have long been held in as much contempt as UFO spotters, not least by politicians and the media. Yet, it should now be obvious that some form of global regulation is rapidly becoming essential. If the growing globalization, encouraged by the worldwide networks of communications, were not enough, one has only to look at the way global financial markets – the first to take maximum advantage of the new communications – allow speculators to outwit national governments and hijack macro-economic strategy.

The first steps are already clearly visible in the form of the regional groupings such as the EU and NAFTA. The former is especially important in terms of the power of the idea of full political union it encapsulates. The latter is important in terms of the moderation it might bring to the external policies of the US, which is seen to be the one state which will move in the opposite direction to the rest of the world into increasing internal anarchy. There are other, hesitant, steps being taken elsewhere towards such supranational groupings, especially in terms of the Pacific Rim countries but also with the Mercosur group in South America. Thus, there is already on the drawing board a three- or four-way split of the world. This will undoubtedly shift as the members of the Third World make their own groupings or join those in the First World, but the principle of large economic – and almost inevitably political – groupings of nations is already established, with a seemingly unstoppable momentum. It should not be an unthinkable further step to reach the ultimate goal of a global society: one which is not totalitarian but which contains a great diversity of approaches among its members. Indeed, it is not inconceivable that world government is already present, in embryo form, as the European Union. The idea, behind what is now a political as well as economic union, is possibly one of the most important shaping the future of the world.

The Political Establishment

Accompanying this social revolution in the position of the individual has been a parallel political revolution, though this has yet to be fully recognized. Politicians in the West have long claimed that democracy, the people's vote, is the best – many of them would claim the only – satisfactory source of government power. It was ultimately this power of the individual, and not capitalism, which defeated the Marxist governments of Eastern Europe. Yet, while claiming to be the guardians of democracy, Western governments themselves have signally failed to recognize the implications which emerge when people like us come to demand that power really *is* distributed to the individual. The lessons of Eastern Europe, graphically conveyed on television, have since been well learned, not just by those who have thrown off the yoke of Soviet imperialism but by all of us in the West. They have been leveraged by the move to genuine equality; not least by the remarkably effective feminist movements. Above all, as we have seen, women have come to realize that they are individuals with real power, not just dependent members of a family, and are now *leading* the revolutions. Individualism has been seen in our buying power, as the conspicuous consumption reflected in our individual lifestyles, and more recently as our responsible buying to protect the green environment. Potentially most important of all, though, it is now being seen in our voter power: our voting out of office, around the world, governments which have long considered that they had a monopoly on power.

At the same time, the spread of power has changed dramatically. For most of the past century it has traditionally belonged to national governments, even where the nation was an artificial construct. Now it is being stretched in two opposing directions. The regions, often 'tribally' or ethnically based, are demanding an increasing say in their own affairs, and below them even separate communities may demand to decide their own future. In the other direction, as globalization bites, governments are increasingly being forced into supranational groupings. These are, for the first time, starting to garner their own powers, which overrule the national powers of their members.

The confusion and uncertainty, caused by these revolutions coming along at the same time, have resulted in a great deal of pain for some of us. These 'revolutionary pains' are an inevitable accompaniment to the changes which are experienced in any significant revolution – let alone in the number of separate revolutions we are now experiencing. Indeed, it is remarkable, in view of the historical precedents, that we are not suffering more pain. It is especially encouraging to note that the social and political revolutions have, in general, been almost bloodless – even that which broke the sway communism held over almost half the world. If this is any evidence, humanity has matured to the point at which we can be trusted to steer our own course to the future.

Even so, it is these pains, the symptoms of revolution, which most of us (and almost all politicians) observe; and it is these on which we focus in our debates about the future. Because we do not look beyond these immediate symptoms to see the underlying, structural changes which will eventually bring great benefits to most of us in the longer term, we adopt policies which make the position worse. The problem is, thus, exacerbated by the short-sightedness of those of us who should know better. These short-term pains are the most immediate problem – and the main threat to a positive outcome in the longer term. The main theme of this book is that we can create a better future; a much better one if we want to. The corollary, though, must be that we can also create a worse one!

Shaping the Future

The first step in creating the optimal future is simply in recognizing the alternative futures which might come about. The technique we use in this context is scenario planning and, accordingly, we have developed a range of approaches which make this much easier to implement. Beyond this, though, there is a need to plan for the longer term; to produce robust strategies which ensure survival. We have found that the problems for many organizations, just as much as for governments, come about because this process is fatally confused with the production of the shorter-term strategies (the 'corporate strategies' beloved of commercial organizations) which aim to optimize current performance. The problem is best illustrated by comparing the two:

	'Corporate' Strategy	'Robust' Strategy
Objectives	Optimizing performance	Ensuring survival
Characteristics	Short-term, Single-focus	Long-term, Divergent-coverage
Outcomes	Effective commitment	Comprehensive understanding
Beneficiaries	Individual profiteers	Community stakeholders

It should be clear, from this table, that the two are potentially in conflict. Fortunately, as we will see, the generic robust strategies, which are already in widespread use, minimize the danger. Even so, it is easy to see why short-termism, perhaps the worst legacy of the 1980s, can be so damaging.

Fortunately, a new mood of co-operation – addressing long-term issues of mutual importance – is emerging as the Millennium approaches, and the presence of such a global consensus is especially important where we can now begin to shape our own destiny. Despite the earnest efforts of academics, not least amongst economists, to find the exact equations which govern the behaviour of society, the outcomes are typically far more chaotic than they allow for. The solution, however, is remarkably simple. Our research shows that if you can persuade the members of society involved to agree not just on a desired outcome but on the expected one, it is likely that is exactly what will happen.

In fact, much of society now looks to 'rational expectations' – to misuse a favourite phrase of the later monetarists – to set their personal expectations of what will actually happen. These are outcomes which they are persuaded will inevitably happen given certain inputs. If enough of us put weight behind the same 'rational expectations', and the inputs are observed to happen, we will adjust our behaviour to take account of what is expected; and, lo and behold, the expected outcomes will occur! The lesson should be clear for all to see. If we can only agree what it should be, the future is ours to shape.

Expectations

The greater part of this book describes the future. At the same time, as we have already seen, it introduces a number of new concepts – which are small in number but remarkably powerful in effect. Thus, the book is based upon three major assumptions:

1 The future of humanity is, in general, no longer constrained by any significant shortage of resources.

2 Accordingly, that future is now being progressively determined by social decisions, taken not just by a few leaders but by millions of us taking billions of small decisions as part of our daily lives.
3 The general, longer-term framework within which these specific, individual decisions are taken is largely provided by our expectations of what the future holds for us.

The result of these three assumptions is two major outcomes:

1 If you can accurately measure these 'expectations' as to future developments to a large extent you can as accurately predict the most likely form of that future.
2 If you can shape these expectations, by whatever political or persuasive processes available to you, you can shape that future and bend it away from the line it is currently following.

These two outcomes lead to one philosophical statement: We all have the right, and duty, positively to shape our own future. For the first time in history, we – as a civilization – can now choose our own future. Unconstrained by shortages of resources, we can decide for ourselves what we want for ourselves and for our children. This is a new privilege – but it carries an equivalent responsibility. No longer will we be able to blame others for our history.

Different Views of the Future

This book contains descriptions of the great many expected changes our research groups predicted, and supports these with further references from a wide variety of sources. As such, while it is undeniably comprehensive, it might be confusing for more general readers in terms of the overall patterns of change which lie behind these details. Accordingly, to help organize these ideas into a more immediately meaningful whole, the material is collected into twelve separate chapters – each of which addresses a set of related issues.

On top of this, there is a further framework in the form of four main sections which, between them, contain all the key developments to be expected. These four sections are further grouped into two sections. The first section ('incremental technology', 'society's winners' and 'a shared world') reflects the views of our research groups (and most futurists) focuses on generally *optimistic* outcomes. The fourth scenario ('dark fears') could potentially lead to *pessimistic* outcomes. The overall balance between these scenarios, three 'votes' for optimism to one for pessimism, reflects our own estimate of the most realistic odds on the final out-turn.

Thus, the sections are:

INCREMENTAL TECHNOLOGY: this looks at technological developments in general, including those in medicine along with those concerning the development of outer space and IT-based communications. It reflects the potential already being demonstrated in these fields, the underlying assumption of the book as a whole, that humanity's resources are now effectively unlimited. It is the most clearly defined view of the future, as well as the most optimistic. It, along with that of 'a shared world', is incremental in form – the changes will come about gradually – where the other two ('society's winners' and 'dark fears') may be subject to major discontinuities.

SOCIETY'S WINNERS, as indicated above, is a much less certain view, reflecting the, as yet, unknown shape of the society which will finally emerge from the moves to individualism – post-modern and/or post-material – and to new forms of community based upon more feminine values. It, too, is an optimistic view, which mirrors the very real benefits arising from the empowerment of the individual, especially of women.

A SHARED WORLD again reflects a more predictable (incremental) development as the large populations in the Third World come to receive their fair share of global resources and ultimately to dominate worldwide economic and political processes. It is an aspect of the future not yet understood by politicians; just as misunderstood are the accompanying moves to power sharing at different levels – from the community up to supranational groupings but always away from nation-states.

DARK FEARS: this view will probably not determine the long-term future, but will determine the degree of revolutionary pain we will experience on the way to that future. It describes the fragmentation of political certainties and the – futile – struggles of the establishment to retain power against the emerging forces of change.

Now, please turn the page and step into the future!

2

INCREMENTAL TECHNOLOGY

This first part reports what is, in many respects, the continuation of our present progress: the incremental progress of science and technology – making us ever wealthier and giving us an ever better qualify of life. Because its progress typically represents a clear extrapolation of current trends, this is the aspect on which most writers about the future have focused. As a result it is the one with which we have become most familiar; indeed, it is probably the only one we recognize.

The pace of such change will grow ever faster, but as the theory behind many the key drivers of technology over the next few decades has already been identified – and tested in the laboratories – though not yet implemented, its overall course is relatively predictable. Furthermore, even the increasing rate of change no longer poses major problems for us. We have become accustomed to it and it has become a fundamental tenet of our society. There are, however, a number of potentially uncomfortable discontinuities still to be resolved as a result of developments in this field.

Unlimited Resources

Not least of these is the fact that we already have the resources available to carry out most of the activities needed to secure our future. For the first time, we now have unlimited resources at our command. Rather, I should say, society as a whole has; for you may still feel, as I do, that you could individually do with considerably more personal wealth! The limitations on change now must be found in the forces described in the other chapters, those which explore the management of society and of the changes within it. Most of the major discontinuities are therefore described later.

One potentially massive discontinuity still remains and that is the development of a new form of humankind, '*homo integrans*', which will emerge from a synthesis between medical developments, discussed in this chapter, and IT developments, discussed in the fourth chapter. If we are correct, and all the evidence supports our predictions, over the next few decades you will become a markedly different person: one whose identity progressively – albeit almost imperceptibly – merges the traditional self with a new one held on computers around the world. Where each of us begins and ends will become progressively blurred, offering a form of immortality!

The technologies described are already in existence, albeit in embryo form, and the only real uncertainties are how and when we will implement them. We should also ask whether they will be implemented. In general, therefore, the next three chapters are an extrapolation of humanity's exponentially growing exploration of, and exploitation of, all aspects of science and technology. In the best of all worlds, these will be our servants; they offer a picture which is highly optimistic about the future.

Optimism and Boom

One of the earliest elements of our quantified research (in 1993, in the middle of the recession) showed that, despite the overall pessimism of the politicians and media, managers were even then clearly optimistic about the future. On a scale of 0 (pessimism) to 10 (optimism), the mean figure from that earlier research was 5.5, which was already close to the ideal position for a happy future, showing general optimism, but not unduly so (since over-optimism might easily collapse in the face of the first setback). One problem, however, was that while managers were themselves optimistic, they believed that most other people's views were pessimistic. The true underlying optimism was, therefore, often hidden beneath a veneer of pessimism. Whether we base our actions in practice on optimism – our own feeling – or on pessimism – what we think others feel – may be critical.

As confidence has returned to the economy, the levels of optimism have also risen. The comparable level in (mid) 1997, measured by a more direct technique (and specifically asking about 30 years ahead), was 6.0. The role of the media is probably crucial in this context. There is a well-reported sociological phenomenon, called the 'amplification spiral', whereby stories reported in the media (whether true or false, but usually controversial) produce a response in the readers which is then fed back to the media again – to amplify what has been said by the original reporters. They, in turn, feel more confident of their position, inflating their reports even further, and the spiral grows!

However, in its extreme version it becomes 'moral panic', where false reports are blown up out of all proportion to create, literally, a form of panic among the population at large. The widespread reports of Soviet troops landing in the UK during World War I, evidenced by soldiers with snow on their boots, is one comic example. The US witch-hunt for communists during the 1950s was, however, a more sinister one. It is just possible that we are once more reaching this point of moral panic in terms of undue pessimism about the future (Celente, 1997).

The signs of recovery are already to be seen around us. Fortunately though, we seemingly cannot believe our luck – and suspect the recovery is fragile.

There is another hidden feature. A hidden, but very important, assumption made by all our research groups was continuing growth. They assumed, almost without question, that the world economy would continue growing. The equally hidden implication is that, in the case of individuals' own decisions – which accumulate across the whole population to set the tone for the entire economy – these decisions are likely to be based on some degree of optimism. This is even the case where they think, probably correctly, that at the macro level politicians' decisions are based on a pessimistic analysis of the future. It is the optimism of the individual, fortunately, which will ultimately count.

Beneath the surface, rampant optimism is driving the increasing pace of change. Buckminster Fuller estimated that about 5,000 years ago a significant invention occurred every two hundred years. By AD 1000 the pace had speeded up to one every thirty years, and by the Industrial Revolution it was down to six months. By the end of World War II there was a significant invention every month, and by now it must be down to a week or so – if not just days. Where it does not threaten individuals (and it must be recognized that at times the pace of progress does cause 'revolutionary pains' on a scale sufficient to make people feel threatened), such a rate of change creates a mood of optimism, if not euphoria. Thus, although this chapter revolves around technological factors, at its heart lies a very optimistic view of an ever-better future driven by these scientific developments. Many of the more popular futurists hold similar views of the future; and their continuing popularity is probably ensured by the technological utopias they describe which we all want to hear.

Malthusian Pessimism

On the other hand, it has to be reported that the numerous, highly vocal, followers of Malthus – who are continually expecting the world to run out of resources – would say that the participants in our research (along with many futurists) are living in a fool's paradise. While the Malthusians do have an argument to make, we believe that, as yet, such pessimism is unjustified. The evidence suggests that, despite the doom and gloom in the media reports, key resources are not yet on the point of running out (Avery, 1995). Expansion can continue, for a while anyway. We can all breathe more easily, perhaps literally so if some of the extreme views of the environmental pessimists are confounded.

Indeed, this expansion may be quite literally the case. In the longer term, salvation may well come from migration into space. Paul Kennedy (1993) makes the important historical point that one major reason why Malthus was proved wrong was the massive migrations which took place in the nineteenth century. This is a rarely quoted reason for the Malthusian trap having been averted. With no new lands on Earth to migrate to, migration has not been an option for solving similar problems in recent times, but by the end of the twenty-first century there are likely to be comparable migrations to other planets and to space colonies. This should please both the optimists, who want to see such expansion, and the pessimists, who will seize on it as the final proof of the Earth's limited capacity.

Millennium

One final factor, to add to the mixture, is the impact of the Millennium itself. It may only be a psychological element, and it certainly is illogical, but the indications are that it, too, will eventually give a boost to the economy. So many companies are once more looking to the future, possibly for the first time in more than a decade, and investing for the long-term potential they will see in the new century. In this, they too are likely to be influenced by the media, which have exhausted the news value of Armageddon and are now on the bandwagon celebrating the new horizons opening up.

To summarize what our research groups decided on this issues:

The importance of the availability of unlimited resources cannot be overestimated. So far, the onward march of humanity has been limited, at every step, by the resources available to it; and trade-offs, often painful ones, had to

be made between conflicting options. In general, this need no longer be the case. Within very broad limits, we now have the abilities, and the resources to match these, to do almost whatever we want. Now, at long last, we are also regaining the optimism necessary to make full use of these. Our future should be characterized by ever-expanding horizons, no longer by conflict over scarce resources.

Beyond Affluence

It is now four decades since Kenneth Galbraith published his famous book *The Affluent Society* in 1958, but few of us learned the lesson it preached: that the economics of an affluent society are very different from those of previous times when resources were rationed. Now, more than a generation later, we have already gone beyond mere affluence – certainly in the West, and the rest of the world probably is not too far behind.

> What makes the Third Wave economy revolutionary is the fact that while land, labour, raw materials and perhaps even capital can be regarded as finite resources, knowledge is for all intents inexhaustible . . . knowledge can be used by two companies at the same time. And they can use it to generate still more knowledge. Thus, Second Wave economic theories based on finite, exhaustible inputs are inapplicable to Third Wave economies.
>
> Alvin and Heidi Toffler (1994)

This is precisely the assumption – of effectively unlimited resources – on which much of this book is based, though there is now so much evidence to support this view that it should really be seen as an established fact.

On the other hand, it should be reported that almost all our groups still expected there to be some problems with food supplies, in terms of shortages – perhaps created by plant diseases, pests or pollution. And, in terms of the results from individual respondents (which better indicate the current expectations), 60 per cent foresaw a global shortage by 2030. Just over half the groups also foresaw shortages of other non-renewable resources. Just over a third highlighted the specific problem of water supplies possibly giving water utilities considerable political power and possibly even leading to 'water wars' (expected by 80 per cent of individuals) where the resources crossed national boundaries. Our contacts with the major multi-nationals working in this field suggest, however, that the problem is typically not one of absolute shortage but of pricing, especially in terms of the massive subsidies on irrigation water (in California just as much as in Saudi Arabia) which then result in unsustainable demands on the aquifers supplying these.

Assuming that resources are effectively infinite is necessarily a crude model, for we cannot instantaneously meet every demand which appears. The point is that even this crude model is preferable to the one currently in use which emphasizes our limitations, often artificially induced ones (such as money supply) in the context of the longer term, and demands that we must fight each other for these scarce resources.

The Green Future – Sex, Drugs and Rock & Roll?

An important, but rarely recognized, addition to this debate is that resources should be measured relative to the likely demands upon them. Most Malthusians assume that our ever-growing wealth will pose ever-growing demands on our physical resources, as has happened in the period since the first Industrial Revolution. On the other hand, it is already clear that the post-modern society (and especially the post-materialist one) will require progressively less in the way of additional physical resources. By definition, post-materialism is not based on the use of materials. The demands are, instead, made on intangible resources; usually provided by human beings, or computers, who are not yet in short supply. The relative balance between resource supply, which is still growing as fast as ever, and demand, which will not grow as rapidly as it has in the past, should therefore improve.

Indeed, in the fast-emerging knowledge society, not merely does use of the 'product' being consumed not reduce the overall resources, it may even increase them. *The Economist* (20 December 1994) puts the point succinctly. 'Economic theory has a problem with knowledge: it seems to defy the basic economic principle of scarcity. Knowledge is not scarce in the traditional sense – the more you use it and pass it on, the more it proliferates . . . however much it is used, it does not get used up.' To put it another way, David Ricardo's widely accepted concept (postulated in the early nineteenth century), the notion of diminishing returns (on capital) which is built into much of (neo-classical) economic theory, no longer holds true.

At a recent meeting of futurists, discussing the best route to a 'greener future', one of the participants bemoaned the fact that all his children wanted was 'Sex, Drugs and Rock & Roll'. This was clearly said for comic effect. But the debate it sparked off showed that, whether or not you agree with the moral values implied (and this is just one of many areas where society's formal position is under challenge), such a range of 'activities' actually offered a remarkably green future; for none of the three activities made any significant demands on the physical resources available to society!

Affluence poses different challenges – of a political rather than of an economic nature.

Medicine

Medicine has an especially important impact on all of us as individuals. We are all, to a varying degree, hypochondriacs! In any case, the rate at which scientific frontiers are being pushed back is perhaps most evident in the field of medicine. It should be emphasized that, in general, it is in this wider field, rather than the more glamorous one of surgery, that the key advances are now being made. Even so, advances in surgery – probably combined with new medical treatments – did feature in the predictions of our groups in the context of brain surgery. More than half our general research groups suggested that brain transplants or implants would be possible (including those with microchips to enhance the brain-power of the recipient, especially in terms of knowledge). The brain is the ultimate transplant; you could, in theory, 'grow' a new young body to house your old tired brain. On the other hand, as we will see later – in the chapter on IT – symbiotic implants are already in the pipeline, one of the key 'technologies under development'.

At a slightly more mundane level, the use of advanced medical imaging and robotics will revolutionize even routine surgical procedures. Thus, *The Economist* (9 March 1994) points out that 'By hooking up conventional medical tools to computers, image-guided therapy offers surgeons the ability to see into and through patient's bodies.' The magazine adds, rather brutally but accurately, 'Surgeons should then do less damage to their patients!' The future might be even more revolutionary, for *The Economist* explains that 'Mini-robots are another possibility. Electric motors less than a millimetre in size, made by etching tiny gears and mechanisms on a silicon crystal, could power tiny surgical devices or tractors bearing cameras. Such miniature battalions can be swallowed . . . eliminating the need for invasive surgery.' John Petersen (1994), describing one example of this, writes that: 'Minuscule devices smaller than red blood cells could cruise that bloodstream searching for fat deposits and infectious organisms. When they find them they would destroy them.' Mind you, this is what white blood cells already do, usually to great effect, so the makers of these tiny machines will here be in competition with the biotechnology industry, which is approaching the problems from a different direction; but, as with many aspects of technology, such a multiplicity of approaches may just make the solution that much more certain and offer all of us a healthier future.

Medical Treatment

It is arguable that, for much of the nineteenth century and some of the twentieth, the improvement in general living conditions, rather than surgery (or even medicine), was the main contributor to improved health. Cleaner water and better sanitation probably saved more lives than all the doctors put together. Now, though, doctors are conquering many of the infectious diseases which were the traditional killers. Indeed, in view of the size of the challenge, it comes as somewhat of a surprise to learn that, as Chris Mahill reports (*The Guardian*, 2 May 1995), 'eight out of 10 children in the world have been vaccinated against the five major killer diseases of childhood [and] since 1980 infant mortality has fallen by 25 per cent while overall life expectancy has increased by four years to about 65.' The problem is, unfortunately, the distribution of this new robust health which the world is experiencing. Again as Chris Mahill reports, 'A person in the least developed countries of the world has a life expectancy of 43 years.' The difference is not a function of medicine but of political decisions.

Thus, smallpox has already gone, and polio may soon follow; though malaria is proving very stubborn and, of course, new diseases – such as AIDS and CJD – continue to emerge, though 90 per cent of our individuals thought there would be a cure for AIDS within the next two decades. With better international controls, however, the impact of these new diseases on individuals is likely to be much less than the alarmists would suggest. We just lack the will (and consequently the dedicated resources) to offer these solutions to everyone, around the world.

Increasingly we are even learning to deal with the problems which occur, at one extreme, within the human cell, and, at the other, in the holistic environment to which the individual is exposed. Once again, more than half our research groups thought that (all) diseases would be eradicated or at least controlled. Perhaps surprisingly in view of the media hype, only just over a quarter of the groups thought that cancer would be cured or prevented.

As in much of the rest of human life, however, here too it seems that IT may have a major – albeit less obvious – impact. Where cost is becoming a major, perhaps (in the form of cost effectiveness) *the* major, factor in treatment decisions – we can now cure more things than we can afford – the use of computer controls on administration may prove to be one of the major drivers for better, or at least affordable, medicine. At the same time, better communications will ensure that best practice is more widely disseminated. To put it bluntly, your doctor will have less excuse for allowing you to die simply because he or she didn't realize there was another alternative, a not uncommon occurrence even now.

The most important developments are likely to be seen in the management of health. The UK government's Technology Foresight Steering Group, reporting in 1995, made the point that 'In 15 years time the cost of treatment rather than a lack of knowledge will probably be the chief limiting factor in the treatment of today's currently incurable diseases – if that is so, science and technology will be increasingly targeted at reducing costs.' Once again, technology will not be the limiting factor. As for so many areas investigated in this book, what we do will be decided by political decisions.

Brain implants are likely to become a possibility, though the timescales for such procedures to be generally adopted are uncertain, and should probably be seen as long-term but very important developments for the future of humankind.

Medicine is, given the resources, already meeting its traditional challenges. Infectious diseases, and those of poverty, are being conquered; and the everyday problems of patients are being addressed more and more successfully.

Longevity

Perhaps the most direct evidence of the medical advances described above will now be seen in the increased life expectancy to be experienced by almost all populations. This will be seen in terms of the absolute length (not just a reduction in infant mortality), especially in the developed countries, as doctors come to understand the mechanisms which lie behind the various diseases, and behind the ageing process itself. Two-thirds of our groups were expecting significant increases in longevity; no less than two-thirds of individuals suggested that this would reach a hundred years or more on average and that this could be the case within four decades.

As already explained, the earlier stages of this process can be largely explained by improvements in our environment. It is only in recent decades – typically in developed countries – that improvements in health care have significantly entered into the equation. Even then it has probably been the breadth (the quantity) of its provision rather than the individual offerings (the quality) which has counted most. In fact, despite the widespread belief in the rationality of medical science, most of the discoveries, especially in pharmaceuticals, have so far come from trial and error – albeit on a grand scale, costing hundreds of millions of dollars. The greatest invention of Edison has turned out not to be the electric light but the modern process of research and development (R&D), which he invented in order to find the best filament material for the lamp.

In the near future we may see average life expectancy moving steadily towards a hundred years or more.

Cell-Level Medicine

All our general groups rated genetic engineering as a major driver for change. This is not to say that we so far have many answers. The human genome project will, by itself, not provide the solution to all the ills of humanity, as some of its supporters would suggest. Its real benefits will accrue from the foundation of knowledge it provides for later work. Even so, we should not expect to see the wave of practical, medical advances it promises put into widespread practice until the middle of the twenty-first century, though this will be when many of those already born will need them. It is an important project, though, when it is remembered that it was only in 1972 that the first gene was cloned. With, or without, the knowledge it provides, we are developing ever more advanced capabilities to repair the human body at the cellular level. One way or another, we can now, or soon will, send biological or molecular (or even electro-mechanical) devices into cells to diagnose faults, followed by others to repair them. It allows the application of at least some of the genuinely scientific method, which has been used so successfully in other fields, to the treatment of many more illnesses.

One outcome of this process is that health in general should improve. Not merely will life-threatening illnesses, or those which severely incapacitate, be deemed suitable for treatment, but many of the things which reduce the performance, or quality of life, of an individual will also be considered to be treatable and, as diagnostic tools improve along with treatment methods, affordable.

Cancer and Viruses

On the other hand, many diseases will prove more complex, and more difficult to treat, than we expect. As mentioned earlier, only a quarter of our groups thought that cancer would be cured by the year 2020. There are, though, many different forms of cancer. Some of these, Hodgkins disease and some leukaemias in children, have already proved to be treatable. Some, unfortunately some of the most common, have not – and the mechanisms underlying them are, as yet, poorly understood. There is little evidence that any breakthrough in treatment of all of these is upon us; so again we must assume that, in general, the effects of such spectacular breakthroughs, even if they happen, are unlikely to be observed much before the middle of the twenty-first century – though the number of cancers which are treatable will steadily increase.

Many viruses, especially those reducing the quality of life (such as colds and influenza), are also a mystery – at least in the way that they transform themselves into new varieties so that, just as we have the answer to the last one, a new one comes along! AIDS is, of course, the most-feared example of this, and in one sense we are lucky that its specific means of transmission offers such a weakness that we can control its spread to some degree, and, in any case, more than 90 per cent of individuals expected a cure to be found within twenty years. There are, though, other similar viruses waiting in the wings – so it would be unrealistic, again, to see a magic solution to the overall problem being applied in the first half of the twenty-first century. This is one area where social expectations may not be able to deliver technological solutions!

Genetic Manipulation in Plants and Animals

More than two-thirds of our groups believed that new forms of animals will be created, usually for food. Surprisingly, in view of the actual developments being reported, only one group mentioned their use in producing organs for human transplantation. On the other hand, Pearson and Cochrane (1995) predict that, by 2020, 'Many synthetic body parts [will be] available.' More specifically, they suggest these will include: 'Artificial ears, eyes, legs, lungs, hearts, pancreas, liver, kidneys, blood'. You could almost build a complete human being from these, but their suggestion is not that outrageous, since crude versions of a number of these are already available.

The wider potential is already evidenced by our ability to produce new animals which deliver products that are compatible with those of humans – such as the blood product, Factor VIII, in cows' milk. It has been evidenced longer in the manipulation of plant species which, at least in part, heralded the green revolution, saving the Third World for a generation or so. But it has been around much longer as, for example, in the manipulation of yeast in the brewing industry; and, even more spectacularly, in our manipulation of the wolf's gene pool to produce the Chihuahua at one extreme and the St Bernard at the other. There is nothing new about the principles underlying genetic manipulation. What is new is our ability to produce the changes we seek, on the scale of these and, in particular, on the speed with which we can make them. One outcome of this is that biochemical processes are increasingly taking over from purely chemical ones – on the industrial scale – with consequent improvements in efficiency. As one result, the new chemical factory may well look more like a small brewery than a massive oil refinery.

Human Genetic Manipulation

Clearly, these techniques might be just as easily be extended to human genetics. In some countries, significantly more boys than girls may be born during the first decades of the twenty-first century – following an age-old tradition which has now been given a mechanism for implementation. Some of these may be taller than they might have been, since taller men have more successful careers. It is likely that both of these trends will, fortunately, disappear when the full impact of female power is realized.

It would, in theory, be possible to create super-strong men, or super-intelligent ones, but it is much more questionable whether this will in practice happen by genetic manipulation. For one reason, the interaction of human genes, especially those controlling intelligence, is so complex that it would not be possible within the foreseeable future easily to predict the outcomes. The super-strong might be subject to crippling auto-immune diseases. The super-intelligent might be impotent. Thus, the parents of the baby-to-be are more likely to worry about negative defects than positive advances; and this will result in considerable conservatism, in terms of opposition to any changes, at the level of the individual. The main reason in the longer term may be that the form of symbiosis with IT systems, mentioned in the section on (brain) surgery, may offer a much more dramatic – and much more (safely) predictable – form of human evolution.

One more extreme view of the future would have individuals being cloned. Surprisingly, perhaps, more than two-thirds of our general groups believed that this was a possibility, though they were all vague about what exactly they meant by this. Was it to counter genetic diseases, which does look like a welcome possibility, or was it to have a 'tailor-made population', as one group put it? In view of what is already being done with animals, the latter is technically quite feasible, but it is a very contentious – and indeed sometimes abhorrent – issue, especially in view of the still-remembered experiments of the Third Reich. Just under a half saw 'baby farms' to be a possibility – though not necessarily related to cloning.

In effect, where in other animals it might be largely an economic issue, in humans such genetic engineering is a moral issue – bringing into question exactly what is human. If we really could create a new race of super-humans, how many would want to consign ourselves to the category of second-class citizens? In the short and medium term, I suggest that few, apart from fanatics, would choose to do so. In the longer term, however, it may be an issue with which humanity may be confronted.

Genetic Preventative Medicine

Some engineering at the cellular level may be undertaken – perhaps even involving manipulation of DNA – to remove those elements which clearly limit human lifespan. As a result, the average lifespan may eventually exceed 120 years, which seems to be the current ceiling, with major implications for managing the human life-stages referred to later in this book. It will also significantly change the return on investment in human infra-structures, in education for instance, doubling this from a crude ratio of 2:1 (for a working life, say, of forty years after an education of twenty years) to 4:1 (eighty years working life, on much the same period of education). This will make it that much more attractive to invest in education, even of the underclasses, which will have dramatic social impacts as well as economic ones.

The human genome project will take much longer to deliver real benefits than is promised, but the processes it embodies will move medicine into a new era and will deliver levels of health previously unrealized. The use of genetic engineering, in as yet unspecified forms, will become widespread in the near future.

Some diseases, which have many forms and especially those which can change over time, will be much more difficult to treat. Thus, some widespread cancers may not be susceptible to effective treatments during the first half of the twenty-first century – though they may be soon thereafter, in time to meet the needs of many of those already born. Viruses, including those similar to AIDS, may be even more difficult to treat but may be susceptible to control of their spread.

Genetic engineering has existed for many centuries. What is new is our ability to interfere in the process much more directly, producing changes on a large scale over short time periods. For the first half of the twenty-first century, it is unlikely that there will be major developments in human genetic engineering – in terms of producing new human beings, rather than repairing existing ones. This will probably not be the result of scientific limitations, but will be on moral grounds based upon self-preservation and self-interest. Even so, some human genetic engineering might extend our average lifespan beyond the 120 years ceiling by the middle of the century. This will have the greatest impact on society.

Drugs for Work and Pleasure

With the growing knowledge of our bodies, and especially of the mechanisms within the brain, we are likely to be able to develop devices – most probably in the form of drugs – which alter the way our brain operates at any particular time. These may offer us the prospect of being more productive at work; enhancing our learning capabilities, say, or just our ability to get on with the more annoying of our fellow workers – as first Diazepam and then Prozac have already offered help to the more stressed out. Pearson and Cochrane at British Telecom (BT) make this point (1995) when they say that, by 2020, 'Brain and mind manipulation will allow control of emotions, learning, senses, memory and other psychological phenomena'. Such drugs usage may, in any case, become a necessity where, according to researchers from BT Laboratories, 'we have reached our maximum information-processing capacity, or at least are within 20 per cent of it' (Ward, 1997). Under these circumstances, it seems inevitable that such drug usage, at work, will grow in scale.

Above all, though, these may offer us new pleasures: heightening our senses, as Ecstasy does, or dulling them, as heroin often does. The problem, again, is probably not technical but social and moral. Interestingly, our individual responses showed, contrary to the reported ethical stance of society as a whole, nearly two-thirds expecting drug use to be widespread (by 2015), to be legalized (by 2020) and to be used for work and pleasure (by 2025). So our formal position, as society as a whole, is looking increasingly out of step with the reality on the ground. It is likely, therefore, that we will ultimately have to review even our so-far universal rejection of the leisure uses of all such mind-altering drugs. In view of the changing nature of our society, coupled with the attraction of such drugs and the growing use of them by large sections of the population, it seems likely that society's view will change and their use will at least be decriminalized.

More powerful, but safer, mind-altering drugs are likely to be developed – and legalized – for use at work and, especially, at play. The main question will then be how we are to best use them.

New Technologies

We now move into the areas which will have the greatest impacts on the economic resources available to us. Technological development has, to date, largely been evidenced by our ability to do much the same things as we have in the past but more efficiently. We are regularly finding ever more efficient, and effective, ways of using the physical resources available. Micro-miniaturization has, across a wide range of fields, reduced the actual amount of physical resources needed to carry out a given function; and this, coupled with the growing amount of computing power built into even the simplest devices, has made these functions ever more efficient in their use of energy. The average family car now contains more computing power than the moon-landers of the Apollo mission. This means that it can travel significantly further on each gallon of fuel, at the same time as offering a more sophisticated driving environment.

As a result, we have been able to afford more, of both products and services. This trend will continue in terms of the basic technologies – even those not directly affected by advances in IT. For the steady advance of all aspects of technology over the past half century has cumulatively changed our lives at least as much. Thus, for instance, the food we eat is fresher and more varied because the container revolution has made transport across the globe affordable.

Technology will continue to develop rapidly, on almost all fronts. More importantly, in the context of this section, we have already invented the new technologies to deliver such incremental changes in most fields over the period – say fifty years – this book covers. Thus, Peter Hall – Professor of Planning at University College, London – says that 'Most of the inventions which will shape the twenty-first century have, almost certainly, already been invented,' but he does add 'The trouble is we don't know what use will be made of these innovations.' What is more, in most fields we are investing in R&D at such a rate that radical inventions might be expected every decade or so. Nathan Rosenberg of Stanford University's Center for Economic Policy comments, on the other hand, that 'What is surprising is the firm's inability to anticipate future applications of their successful innovations.'

It should be emphasized, though, that these developments will be as a result of the development of technology. Its path and that of science, which has supposedly driven technology for the past century or so, will to a degree diverge – with science developing a different existence, one which is less central to the development of society, and a justification of this in the form of the knowledge revolution. The major new technologies, or at least the ones in which advances are already under way, are listed by the experts to be: IT, biotechnology, space travel and (see below) materials.

Materials

You might think that there would be little to be changed in the materials from which our products are made, but radical new forms of these, such as engineering ceramics, will become available to improve the performance of products in general. The greatest impact, however, may come when these are wedded to micro-miniaturized circuitry. For example, Ivan Amato writing in *New Scientist* (15 October 1994) suggests that the new materials, in conjunction with IT networks, will allow city structures, from buildings to bridges, to diagnose their own structural weaknesses and then repair any faults by themselves. Joseph Coates (1996) puts this in a wider context by suggesting that 'Everything will be smart – that is, responsive to its external or internal environment. This will be achieved either by embedding micro-processors and associated sensors in physical devices or by creating materials that are responsive to physical variables such as light, heat, noise, odors and electromagnetic fields.'

It should be noted, however, that as Thomas Eagar (1995) reports: 'There has typically been a 20-year interval between invention and widespread adoption of new material.' This is not just a problem of technological development but of design. 'Product designers tend to use materials in the same ways as the old materials. As a result, early designs with new materials rarely demonstrate their full potential.' The time it takes to get any new invention into widespread use is the main drag on the speed of technological change. On the other hand, it means that future technological developments are that much easier to predict; they can usually be observed already working their way through the system.

Production

Apart from the computer/communications aspects, there is little indication of major new technological initiatives in manufacturing industry. The computer already allows us to do many things in very different ways, but much the same principles have been applied on a considerably smaller scale to, quite literally, operate at the atomic level. In this context, the key image came when – as reported by David Bradley in the *New Scientist* (29 April 1995): 'the letters IBM [were] spelt out on a sheet of noble metal in just thirty or so xenon atoms.' This is one area where the future technology can already be clearly seen. 'Small is beautiful' turns out to be a remarkably powerful concept in design, which was inconceivable a few decades ago when you judged the inherent quality of a product by how heavy it felt in your hand.

It has now been recognized that the key factors in developing quality, productivity and all the other aspects of process improvement come at the stage of product design. Thus, the most important introduction of computers has come at this design stage where Computer Aided Design (CAD) can cut time and effort by up to 90 per cent. In addition, the resulting plans can be very easily modified, removing one excuse for avoiding change, as well as almost instantaneously transmitted to remote suppliers and to computer-controlled machines on the shop-floor.

Even the traditional 'job-shop', the home of 'specials' and one-off engineering and usually full of filth and noise, cannot escape these changes. Rapid prototyping technologies (or FFF – Free Form Fabrication) can build up complex three-dimensional parts direct from CAD input without any other machinery needed.

Computers in Products

To get some idea how micro-miniaturization has already changed your life you only have to pull out your pocket electronic organizer, for the equivalent in the 1960s occupied a whole building and cost millions of dollars.

IT, biotechnology and space travel are areas likely to experience major developments, as is that of new materials. Technology in general, as increasingly different from science, will continue to develop and further enrich people's lives.

Physical Transport and Travel

Our research also indicated that transport in general, and personal travel in particular, will continue to experience rapid, and significant, developments. Along with other developments in IT and communications, virtual travel – seeing the world literally from the comfort of your own armchair – may be one of the most spectacular developments in the coming decades, but the importance of developments in physical transport have arguably been almost as important as electronic ones. The 747 airliner and mass air transportation (for fresh fruit as well as for people), along with the container revolution at sea, have demonstrably enlarged our horizons.

In the short term, though, our continuing struggle through the traffic to our place of work, dominated by what many would call the curse of the automobile, will also loom large. It is likely that this problem will be addressed from two directions. Public transport, improved and subsidized in a political climate which more realistically recognizes the role of government in such matters, will carry increasing numbers of us to and fro and, in particular, will absorb large amounts of freight, to make extra space on the inevitably overused road systems. As indicated, however, these changes will require new policies and the adoption of new attitudes by the general public.

From the other direction, our car will itself effectively become part of a highly regulated public transport system, on oversubscribed routes at least. One aspect of this will almost inevitably be that we will have to pay for such travel. Free road use will, as space on them becomes increasingly inadequate, go the way of cheap petrol. One likely technical development is the ability to run safely our (automated/computerized) cars in very close proximity along dedicated highways – allowing more traffic on the same roads, together with the additional benefit of as much as a 30 per cent saving in fuel.

In the shorter term, however, the problem of car engine emissions may be seen as a higher priority. Energy efficiency in general, and air pollution in particular, will demand new solutions; and the electric car currently seems to be one of the most suitable solutions, also having major advantages in terms of easier control (and higher reliability) for use on the 'automated roads' described earlier.

Almost two thirds of the groups mentioned some aspect of traffic congestion or control of this at the local level, and 85 per cent of individuals expected this to happen within twenty years. At the other end of the spectrum, supersonic air-travel will become widespread. If economies of scale, backed by new technologies, can bring the costs down, the benefits would be spectacular! Breakfast in London, lunch in New York would be opened up to a mass market.

As a footnote, it is an interesting insight, into the investment barriers which can still be brought to bear against new entrants, that – as Paul Krugman (1994a) points out: 'the Boeing 747 [which is] still the flagship of airlines, was introduced in 1969; today's versions are improved, but not radically different.' Even now, its potential replacement, the new Airbus development, has not left the drawing board. Monopolies often work against the public interest not in terms of high prices, as most people expect, but as low investment in key developments. Much the same charge could be levelled against Microsoft in the field of software.

Coming at the issue from a different direction, it should be noted that tourism is already the world's largest industry and – according to Sue Wheat, writing in *Geographical* (April 1994) – 'is growing 23 per cent faster than the overall economy. The number of people now travelling for tourism purposes is around 500 million and is expected to rise to 937 million by 2010. For an industry which doesn't have to manufacture anything itself and simply exploits ready-made features, these are astounding figures.' As a human activity in its own right it cannot be ignored and as a generator of revenues for the developing world it will also be important. Sue Wheat, however, points out that although it employs more than 127 million people worldwide, 'in developing countries, the local people normally work in menial, seasonal and low-paid positions.'

Transportation, of all kinds, may represent an area where major changes might be expected. Few other major technological advances will have dramatic impacts on society, either in the home or in manufacturing industry.

Tourism will become one of the world's largest industries and a key influence on local 'heritage' industries as well as on the economies of the Third World.

Unlimited Abilities

What will emerge, in almost all fields, is the widespread application of the technologies we already possess. The key, therefore, will be which of these technologies we now choose to deploy in any given situation and what priorities we give to each of them. We already have the technology available to make cars which will last for decades, but consumers have been persuaded to view them as fashion items, so we treat them as semi-disposable. We already have the technology to make better, more high-tech, houses but, again, we consumers (and especially the finance houses providing our loans) don't want these.

The key point here is that in general we already have the technology. What we need, in each case, is the will, the consumer demand, to do whatever we want. If we want to irrigate the Sahara, to make it the new Garden of Eden, then this would be possible though it might be prohibitively expensive. The one major exception to this rule of unlimited resources may be that of energy, or at least of *cheap* energy. More than two-thirds of our groups saw traditional energy supplies running out; and about half highlighted this in terms of the 'end of fossil fuels'. This is unfortunate, since energy can be traded off against almost any other commodity or activity.

The need, now, is for rational planning of energy policy worldwide. One 'wild card' speculation by a consortium of three research and consulting firms (The Copenhagen Institute for Futures, Denmark, the Institute for the Future, United States, and Bipe Conseil, France) is that most nations around the world will eventually (by 2027, according to these projections) impose punitive carbon taxes (at $100 a barrel of oil) which will completely change energy usage patterns. On the other hand, as Hamish McRae (*Independent on Sunday*, 26 February 1995) adds: 'At present consumption rates there is enough coal to last the world for more than 200 years,' always assuming that the problems of global warming can be resolved. Energy shortages are not, despite the views of the pessimists, likely to trouble us in the short or medium term.

Ultimately, where electricity will in the long term almost certainly become the main means of transmitting energy to static installations, hydrogen – which has many advantages, not least its zero impact on the environment – may become the major transmission medium for mobile installations (including vehicles). Both of these will, however, be only a means of transmitting power generated from other sources of energy, though this will allow those other sources to be both remote and environmentally friendly.

Nuclear energy, once heralded as a saviour, has now been abandoned by a number of countries, probably unnecessarily, where it is its uncontrolled use which poses problems much as uncontrolled use of thermal power stations may be causing the greenhouse effect. Nuclear energy can be viable, as its use in France to provide the major part of that country's electrical power shows. The problem is the unknown future; the cost, and danger, of dealing with the by-products. Even so, this fear may be less than rational. Two-thirds of our general groups believed that nuclear energy, of one form or another, would be important in future (albeit not until 2030). So it has to be argued that the industry's problems probably are short-term (image) ones. Almost half our general groups believed that fusion energy will be available by 2030.

Even so, we already have large amounts of clean energy available, albeit not cheaply. The highlands of the world contain the potential for vast amounts of hydro-electric power. Just one scheme, dropping the Nile through 6,000 feet and costing a mere $500 million, would produce perhaps a hundred times the power of the Aswan dam (which currently provides a significant proportion of Egypt's needs). The problem is that such power sources (sunlight in the Sahara is another) are usually remote from potential users. The one mentioned above is several hundred miles away from the nearest port, and the subsistence farmers in the area which surrounds it would hardly be able to afford the $500 million capital cost – no matter how cheap the resulting electricity – though it could feed into a transmission 'backbone' already planned to run the length of Africa.

In space, as we will see later, energy from the sun will be relatively cheap. The trick, then, may well be to move the more energy-intensive processes to the supply – on earth as well as in space – reversing the current approach and, once more, demanding planning on a global scale.

We already have the existing technology necessary to achieve almost anything we want to do – it remains for us to decide what we want to do.

Energy will continue to be made available in ever-larger quantities, probably including that from nuclear fusion in addition to nuclear fission. Much of the renewable energy, at least, will nevertheless be found in remote areas – ultimately in space – and almost all of this will be relatively expensive, in the short term at least.

Resource Distribution

As we have already seen, we can reasonably assume that, across the world as a whole, resources and technological capabilities are effectively unlimited, but the distribution of resources is very uneven, especially between the rich and poor and between First and Third Worlds. Such unevenness in the distribution of resources is clearly inefficient from the point of view of the world as a whole. It clearly creates great hardship for those who are deprived, but that is not the result of technology. Nor, though, is it even one of economics – as some would have us believe. It is a political problem. Resolution will require that political decisions need to be taken and these decisions will demand considerable courage from the governments making them – though they probably have no alternative, if the future progress of humanity is not to be imperilled. On the other hand, the World Bank, in its 1991 *World Development Report*, was able to claim, 'Famines disappeared from Western Europe in the mid-1800s, from Eastern Europe in the 1930s, and from Asia in the 1970s.'

The planning of resource distribution, probably applying to new resources rather than to a significant redistribution of existing ones, should be a high priority for any form of world government or of the agencies which stand in for it. This was recognized by the so-called Poverty Summit of 1995. At this summit, President Mitterrand made the very brave suggestion that such redistribution should be funded by a 0.5 per cent tax on every financial deal, a variation on the 'Tobin Tax', which would have also stopped financial market speculation almost in its tracks. It was interesting to note that it required a ruler who was dying of cancer, and had nothing to lose, to suggest such a brave scheme and that almost nobody else gave him any backing! Indeed, the scheme was quietly killed a few months later, a respectable time after he had relinquished the presidency of France.

Wastage of People

It is now clear, at least to well-managed organizations in the developed world, that their most important resource is that of people. The failure to provide everyone with the minimum resources to become productive – let alone to fulfil their ultimate potential – represents the most significant waste of resources around the globe, and the most significant failure of governments of all complexions.

With modern technology, the level of resources needed is now reducing, quite rapidly, rather than increasing as it has in the past. In particular, educational technology – not least that employing distance-teaching techniques – has advanced considerably, so that it may be possible to take a student in the Third World to completion of a degree course for less than $1,000. In addition, large numbers of teachers may be trained by what amounts to a geometrical progression: the first thirty train thirty more each, over six months to a year say, and so on, until by the end of the fourth year nearly half a million students may be taking part. The same geometric pattern may apply to other processes, not least to those requiring small inputs of materials but high inputs of skill.

Third World Potential

Thus, the greatest waste of all is that of the four-fifths of humanity which is locked up in the as yet undeveloped world. This is, paradoxically, where much of the dynamic for development into the later part of the twenty-first century is likely to emerge. With relatively little investment now needed, much of this is set to erupt as those wasted human assets are unlocked. Once they start moving (and they have already started!) their economic growth will be explosive. They obviously need to catch up with the developed world, but South-East Asia has shown how this may be done, against all the supposed odds, in little over a decade. Who would have forecast as little as two decades ago that backward China would be poised to become the world's largest economic superpower in the early part of the twenty-first century.

While overall resources are effectively unlimited globally, the distribution of these resources is very uneven: significant resource constraints still apply in many localities. This is ineffective globally and often literally disastrous locally and should therefore be a high priority for global planning in theory – which it already is – and in practice – which it is not.

The greatest waste of all is that of people's talents – and currently that waste applies to much of humanity. The cost of rectifying this is now affordable within, if geometric progressions are followed, reasonable timescales.

The greatest potential – of resources yet untapped and of the people-power to exploit them – is to be found in the remaining Third World. Following the pattern of succeeding against the economic odds set by South-East Asia, these nations still in the Third World may become the dynamos of the later years of the twenty-first century.

3

OUTER SPACE

One of the most important drivers for change in the latter part of the twenty-first century is likely to be the single decision as to whether we are going to explore, and in particular exploit, outer space. If we decide not to, even by default, then expansion of the territory available to humanity, and development of the associated resources, will be postponed – perhaps indefinitely, as humanity sinks into torpor. It might even signal the end of this phase of our civilization and a return to the dark ages. On the other hand, if we do eventually take the decision to go out into space, and to colonize it, then this will have major impacts – albeit over a longer timespan than many of the other factors reported in the book – even on those of us who remain behind. It will be humanity's greatest adventure: starting with space travel and working up to colonization of the inner planets, and ultimately of the space between them, then probably reaching as far out as the asteroid belt by the end of the twenty-first century. It will, then, signal a genuinely gigantic step forward for humankind.

Thus, among the drivers which were expected to become reality by those participating in the research, this one turned out to be the joker! Since the 'entertainment' spectaculars of the 1960s and 1970s, climaxing in the moon-landings, it has largely been ignored by most of us, apart from a few aficionados and trekkies. Yet, this was clearly recognized by the majority, about two-thirds, of the groups in our research as being one of the major drivers for change for the next decades; and more than half of the individuals expected there to be a moon colony by 2040.

In one form it has, of course, already changed our lives. The many unmanned satellites, which now circle the Earth, allow us, and in particular our computers, to communicate with each other as never before; to obtain television pictures live from the other side of the world, perhaps direct from satellite TV; to see what our weather will be, and in the process to change our image of the Earth for ever; or to sense remotely where there are minerals, or the danger of a drought, or illegal drug cultivation. Already the Global Positioning System (GPS), with a network of twenty-four military navigational satellites, can tell you where you are, with an accuracy of a few tens of metres or so, anywhere on Earth, using just a hand-held receiver. In terms of communications, Motorola is about to launch sixty-six satellites, built by Lockheed, for its Iridium Project so that pocket-sized mobile telephones will be able to communicate anywhere around the globe.

That our interest in space did not seem to survive beyond the 1970s may have been due to two factors. Of these, perhaps the most important was the fact that the prime focus was really that of defence needs – especially putting military satellites into Earth orbit. This military use of space was anything but outward-looking; the satellites were all pointed inwards, surveying the targets, the enemy, below. More, it was backward-looking. If it had anything to say about the future, it was that there probably wasn't any; the rationale for these spy satellites was the destruction of humanity! If nothing else, it meant that all this work was conducted in the utmost secrecy.

The other factor which drove the moon-landings had been the need that US prestige (almost fatally wounded by the achievements of the USSR in Earth orbit) depended upon putting a man on the moon, by whatever means possible. Though this did ultimately seem to put man into space, and onto another body within the solar system, it was in many respects a sleight of hand. It may have restored the image of the US, but at the cost of the serious exploration of space. In fact, it impeded it – making the more prosaic steps towards the real development of space seem boring for most of us watching such events, when we compared them with our memories of the thrill-a-minute which the Apollo programme provided. It even persuaded the scientists that unmanned space vehicles were more cost-effective, which they were in the short term. Neil Armstrong's step was, in retrospect, one backwards for mankind.

Even so, the impact was dramatic in one respect. It was no longer possible to say that humanity could not conquer space. The genie was out of the bottle, and space was there for the taking.

William Braselton – a Vice-President of Harris Government Aerospace Systems – describes space, with some justification, as 'the world's largest industry of the early 21st century', and Vincent Kiernan suggests, in a similar vein, that 'In future, historians may say that the development of such a space infrastructure had the same boost on the world's economy in the early 21st century as did the construction of major road networks this century and the building of rail networks in the last.' I think a better analogy, overall, might be the discovery of, and economic development of, the vast open spaces of America. The boost that gave to the European economy over several centuries, even before the US itself became the richest nation, cannot be underestimated.

All of this was, in one form or anther, recognized by most of our groups and even by a majority of individuals.

The development of space is, therefore, *inevitable.* History demonstrates that nothing will stop a buoyant humanity from expanding its frontiers. That is the key fact that we, and especially our leaders, must recognize. The only thing which can stop humanity expanding into space is that humanity itself has no long-term future! When the large-scale expansion into space will finally happen is, however, much more questionable – though our research evidence suggest that this will be quite soon, at least in terms of cosmic timescales – and it will happen, unless there is nobody left to make it happen.

Our ability to develop space is already a proven fact; it requires no new technologies – only political decisions.

Though it is not yet obvious to outside observers, the pace of development is accelerating once more. Spyros Makridakis, writing at the end of the 1980s, will almost certainly be proved pessimistic when he gives the likely date for the first full-scale space-station as 2030. The first element launch of the joint US and Russian station was originally set for November 1997, and, although the timescale has already slipped, the planned completion date is still 2002. Perhaps Makridakis' definition of a 'full-scale space-station' is more ambitious than this, but I still suspect that the 2030 date will be beaten by some margin.

The development of space to date has been sidelined by short-term military and political considerations. But it will eventually happen, probably starting in the near future, unless humanity decides that it has no future.

Heavenly Resources Unlimited

As with the exploitation of the last 'new world' almost half a millennium ago, the ultimate rational justification for our exploration of, and in particular for our exploitation of, the solar system (and beyond!) will ultimately be the resources made available, along trade routes which will span it. There is every mineral, and every element, we might want out there; often conveniently available – as asteroids, say, with no gravitational penalties – or as carbon reservoirs in the outer atmospheres of the larger planets. It will take considerable effort and massive investment to establish bases from which to develop these resources, but, as the early colonists of America found out almost half a millennium ago, conquering the problems found there will eventually prove to be a worthwhile investment, where it will bring equally massive returns.

Exactly what will be the most important of these resources is difficult to predict, but with Earth running out of easily available supplies of some key minerals, these rare metals may once more head the list of priorities. Or it may be the availability of cheap energy which will become the greatest attraction. Unfurl an aluminized plastic membrane, spin it a bit to give it a reasonably parabolic shape, and you have a giant mirror ready to capture the sun's energy and give you as much power as you want, almost for free!

The long-term justification for space exploration, and especially for exploitation of it, will be the effectively unlimited resources – especially of energy – it makes available to humanity, sufficient for millennia of further progress and expansion.

The Final Frontier

On the other hand, the short-term justifications are likely to be much more *emotional.* Explorers, as opposed to colonists, have almost always been dreamers. They really do want to go, in the immortal words of *Star Trek*, 'where no man (or, in the politically correct version later used, no "person"!) has gone before'. There is a part of humanity which will be desperately unhappy if it cannot explore new frontiers, and there is another large portion of it wanting to cheer them on. It is, thus, probably preferable that the attention of both these groups is diverted to space, rather than to more destructive pursuits on Earth!

Space may, indeed, unlock some of the dynamism of which humanity is capable, literally exploring new frontiers in space (even if it is vicariously through television). It may also defuse some of the tensions which might, otherwise, build up and erupt into violence.

Perhaps the most unexpected outcome of our research was, however, the emergence of space as the ultimate insurance policy; offering a guarantee of future existence for humanity – no matter what disasters hit Earth! Michael Mautner (1994) suggests it could even help avert one impending 'disaster', which we already know about, 'To compensate for the predicted greenhouse effect, we need to screen out about 3 per cent of the of the solar energy incident on Earth. This requires a screen area that is about 2,000 km (1,200 miles) on each side.' The same effect could even be achieved by injecting moon-dust into low Earth-orbits, in effect simulating the large-scale volcanic eruptions which have 'naturally' created the same effect, reducing temperatures around the world, over the ages. Mautner applies the same approach to solar power: 'The basic idea is to collect solar power in space where solar radiation is free of day-night cycles and cloud effects. . . . A few thousand units could provide the entire world's energy needs. This could be the ultimate source of clean, renewable energy.'

Returning to the theme of 'insurance', however, the comment which best encapsulated the concept was made by one of the participants in our scenario-based research, who suggested: 'God chooses a second Ark – a space Ark!' This seems to be a very potent force, and may be the one as to why space will be given some priority. More than two-thirds of our general groups suggested some form of natural threat coming from space itself: from asteroids colliding with Earth to black holes engulfing the Sun. Indeed, the statistics show that (due to the enormous numbers of fatalities which will occur when such an event really happens) we actually are more likely to die in this fashion than in a plane crash! It will also be a force which demands not just exploration or even exploitation, but colonization on the grand scale, reinforcing the emotional appeal of a genuinely new frontier described above.

The most important fact, perhaps, is that we already own the basic, proven, technology for space exploration, and even for its colonization. Whether this technology is deployed depends only upon *political* decisions and, in particular, on the pressures which are brought to bear upon the politicians who will make these decisions. The forces described so far, the final frontier and space insurance, are to a degree positive forces; but more important perhaps, in terms of neutralizing potential opposition, they are unobjectionable ones. There are those few who will demand that funds are diverted from space to their favourite projects, but there will be almost nobody who actively opposes it as a matter of principle. The resulting (relative) lack of controversy will be an especially attractive feature for many politicians. It will represent an area where they can be seen to be active, even daring, without any fear of backlash.

Thus, one potent political reason for focusing on it may be that it diverts attention from problems on Earth, which may seem increasingly insoluble, to space, where they may be increasingly soluble! The two other positive political reasons in favour of the development of space represent the opposite sides of the same coin. One is that of co-operation between nations – as it already is, very symbolically, between the US and Russia in the construction of the first true space station. The other is almost the exact opposite: the competition between nations, as it previously was between the US and the USSR during the Cold War.

Space exploration will represent not just a matter of physical travel, it will also be a matter of group psychology: a cathartic, and productive, release of tensions which might otherwise be deployed destructively.

The most potent but unexpected reason for going into space, among the population as a whole, is that it will provide an insurance for the future of humanity, a very potent driver for large-scale colonization of space.

The exploitation and colonization of space rests not upon untried new technology but only upon political decisions. The immediate reasons may seem to be for co-operation or competition with other nations; but the main reason may be that it shifts the focus away from Earthly problems, with little chance of any backlash!

Space Exploration

The first stage of the genuine development of space is only now about to start in the form of the building of the joint US/Russian space station in the near future, with European and Japanese contributions to this programme also due to begin before the end of the century. The new venture should be the first genuinely productive space platform.

It will still be Earth-oriented, as inevitably will be most of the other developments in the first space-platform era, but it will offer a complete working environment. The great advantage of a manned platform is its total flexibility: missions can be optimized *in situ*, or changed, or new missions added, with little extra effort. The repairs to the Hubble space telescope graphically demonstrated the virtues of direct human intervention. The overhead in building the original platform may be very high, but the marginal cost of adding new missions will be relatively low, an important factor where communications volumes are growing exponentially.

Space Exploration

This is already under way but using unmanned probes, such as Voyager. The systems behind these have already been refined to such an extent that NASA's Solar System Exploration Division can launch planetary missions, for example to land on Mars, for $150 million a time rather than the billions of dollars they each used to cost. Even so, the big difference, in the second stage of development, will be that this exploration will increasingly be by manned expeditions with, once more, a greater degree of flexibility than their unmanned predecessors.

More important, from the point of view of later colonization, will be those expeditions in which a commitment to the long term is clearly made and which are 'staged', putting in place an infrastructure such as orbiting bases around the planets. These will be used by later expeditions and then by colonists before descending to explore those planets and, once there, building permanent bases. The key decisions, here, will be made when these expeditions change from being mere exercises in planting national flags to paving the way for later colonization.

The next phase, albeit later in the twenty-first century, is likely to be a progressive hardening of the lines of physical communication being forged as part of this exploration process. In particular, the (Earth) orbiting platforms will increasingly turn their attention to handling travel, and especially freight transport, to and from the burgeoning colonies. As this transport infrastructure is built, it will probably bring into use the Moon as an important staging post for supplies, at least, and more than half our individual responses expected this to happen by 2040.

The first real steps into space are likely to take the form of Earth-oriented, manned space-platforms. These will be work-sites dealing practically with Earth's problems, especially those of exponentially growing communications rather than space itself. This work has already started, and will probably dominate space activities through the first two decades of the twenty-first century.

The second phase of space will come with manned exploration of the near planets, but especially that which is clearly staged as a precursor for later colonization. This will overlap the first phase, but will probably not come into its own until after the first decade or so of the twenty-first century.

The next phase, perhaps, will be the creation of the interplanetary transport infrastructure. This will again overlap earlier phases but will probably not be consolidated until mid-century.

Colonization

The ultimate goal should be colonization, which may dominate space activities in the second half of the century, though it will come to dominate debate, and the psychology of planning for space, long before then (probably being a major topic by the end of the first decade of the twenty-first century). Indeed, all of the more than two-thirds of our groups who mentioned space saw it in terms of colonization: of the Moon and the other planets, and of space itself. This colonization will probably be comparable in many ways with earlier periods of colonization in Earth's history, involving literally millions of individuals travelling across space to make a new life in these colonies.

Finally, these colonies should become self-sufficient, indeed becoming net exporters. A quarter of our groups specifically mentioned this aspect (albeit not occurring until mid-century). They would then become members of a federation of equal partners within the solar system. Many of these colonies will be on planets, or similar bodies (such as the larger asteroids), and will look like more fragile equivalents of their Earthly counterparts, probably being safely protected underground, with just a few small air-tight domes, until the planets can be terra-formed. The main change from earlier waves of colonization will be that many more colonies are likely to be in deep space itself. Despite the strangeness of the concept, no less than 40 per cent of individuals expected this to happen by mid-century. If you have to maintain an artificial environment, you might as well do this in space where there are the advantages of limitless energy; and the colony can be set in an orbit to cycle between its main sources of supply, or even to operate as an exotic caravanserai for travellers *en route* to the other colonies. The development of these pure space colonies (once more possibly containing millions of inhabitants) will represent a quantum leap in the development of humanity (O'Neill, 1981). Indeed, the first fully productive external colonies might be in space itself, mining the earth-approaching asteroids.

The final goal – in the second half of the century – will be self-sufficient colonies, with millions of inhabitants, as equal members of a solar system federation.

Oceans of Space

Following the theme of colonization of hostile environments – but this time on Earth – almost half the groups returned to a popular theme from several decades ago: colonies under, or on, the sea (by 2040). As nearly three-quarters of the planet comprises such oceans, and half of the mass of living matter lives in them, it does seem inevitable that humankind will eventually put them to better uses than the hunter-gatherer approach now adopted by the fishery industry.

E.T. Phone Earth!

Another surprising result of our research is how many groups included some form of contact with aliens in their scenarios. It is possible to view the likelihood of this event from two extremes, both of which explain the current lack of hard evidence for life elsewhere. At one extreme, the odds against repeating the complex processes which led to the emergence of man are so high that it is unlikely that there is anyone else; or if there is, they are so far away that it may be millions of years before we meet them. The other extreme is that, out of four billion years of Earth's existence, *Homo sapiens* has only emerged over the past couple of million years, and has only developed the technology to communicate across space over the past one hundred years; so if any alien civilization was only 1 per cent faster along this road, they would be forty million years further into their civilization, and as such would probably be totally unrecognizable to us. That brief flash of light you just saw out of the corner of your eye might have been such an alien strolling through from another dimension!

From the point of any scenarios we might want to construct, the outcome is the same, whichever route we follow. We must ignore the impact of aliens until the time, if ever, it actually happens! If it does come, and maybe they are just waiting out there for us to become mature enough as a species to cope with the shock to our pride, it will have monumental consequences but we cannot plan for it.

4

THE COMMUNICATIONS REVOLUTION

The final part of this technological section of the book is concerned with the key issue of what used to be known as the IT Revolution, though it is now increasingly referred to as the 'Information Society'. The technological developments described in this chapter will lead to massive changes in the way we work and live and perhaps even in who, and what, we are. Many of the long-term effects are by now almost inescapable. How they will affect our medium-term prospects, and especially those in the short term, will be a function of how society in general, and its leaders in particular, react to their impacts.

Though the term is still often used, 'IT Revolution' is, in many respects a misnomer since it does not do full justice to the most far-reaching developments taking place around us. Certainly, the massive increases in our ability to store and manipulate data, using ever more powerful computers, will be little short of revolutionary, but even more important will be the new communications abilities, not least in terms of making the massive amounts of information that the various computers around the world will be holding available to everyone, everywhere. They will be especially important in terms of improving person-to-person communication. This has major implications, not least for the 80 per cent of humanity still living in the largely pre-industrial Third World. They may be able to jump direct to the new post-industrial society, missing many of the traumas their Western counterparts had to endure, especially in terms of mass urbanization with which industrialization has, so far, always been associated. I have to admit, though, that despite having explored this concept with a number of leaders in the Third World, I cannot say that the outcome is clear-cut. They would obviously like to avoid the problems of slum cities, but are not sure they can.

In view of its widely acknowledged importance, one might have expected it to be universally recognized as one of the key drivers for change, but the topic received less attention by the participants than one might have expected across almost all forms of our research, even among those actually employed as IT professionals. We suspect that this may have been a saturation effect, brought on by the barrage of media stories. Even so, it is an important finding. If people are becoming apathetic about IT, for whatever reasons, this may eventually reduce its impact on society.

The ways it is generally – and popularly – supposed to be working are usually somewhat wide of the mark. Thus, this revolution is most often described in terms of hardware. This is, of course, important; without it the revolution could not have happened. With the number of computers multiplying by 100 to 1,000 every decade and their power (in terms of whatever you choose: main storage, mass storage, communications bandwidth) increasing by a similar amount, we can foresee computer power being available to handle anything we might dream of. Indeed the much-quoted Moore's Law states that the circuit density of computer chips, and hence their computing power, is doubling on average every eighteen months. Pearson and Cochrane (1995) predict that: 'Given this rate of progress, we can expect to see somewhere between a thousand- and a million-fold increase in electronic capability over the next 25 years.' As these predictions come from workers at one of the leading-edge research laboratories (British Telecom) in the field, they must be taken seriously.

Perhaps important new changes will emerge in one aspect of the technology: the interface with the user. The current personal computer is an awkward mix of typewriter and television, the two technologies from which it has emerged. People can work with it to great effect, but their relationship to the underlying technology, and especially to the information it accesses, could be much improved. Nearly all the groups expected all communications to be 'mobile', rather than tied to the office, within twenty years.

Maybe the better interface will be simply a matter of improving the current workstation configurations with larger, thin (solid-state), colour screens (and more of them) to provide more, and better presented, information. For instance, based on technology already in the laboratories, Pearson and Cochrane (1995) predict that, by 2020, there will be 'large, wall-hung high-definition colour displays' and even 'video walls, including living area use of virtual reality (scenes)', as well as three-dimensional television. Maybe we need a better way of entering data, by voice perhaps, as latest developments already available for ordinary PCs suggest.

On the other hand, the consensus of opinion, amongst the leaders of the IT industry with whom we have conferred, is that the most urgent need is for some form of *artificial intelligence* to be used to provide a more interactive interface with the user. They see this as the greatest failure of the computer industry in recent years and is their main reason for claiming that Microsoft – with its fixation on icons and windows – has held back development for a decade or more! Perhaps, in a few decades' time, our input will be taken direct from the brain's electrical activity. Almost two-thirds of our individuals suggested that microchips will be implanted in the brain (by 2035), in a form of symbiosis, when we will all become superhuman in our responses at least. This will be true virtual reality and, not least, it might allow us to develop a symbiotic relationship with our computers. Eventually it might even give us a new mode of human communication – brain-to-brain – and make us all, in effect, telepaths.

Many science fiction writers have dreamed of combinations of man and machine which give the recipients superhuman strength; few have made the much more important link to combined, superhuman, *intelligence*. This would be a giant leap forwards: no longer

Homo sapiens but *Homo integrans.* It is no longer science fiction, it is (in, quite well developed theory at least) fact – though, once again, it is worth reiterating that it will take a number of decades before it comes into widespread use. Actually, a more limited form of 'symbiosis' is already becoming a fact, though in a less obvious way. We are even now allowing ourselves to be linked to the computer networks. We may not yet have microchips implanted in our brains, but we are increasingly consenting to be tied to our PCs for long hours every day. This link may be much cruder, a display screen and a keyboard, but it will become ever more influential as we learn what the true power of the computer can bring to us. Increasingly, we are becoming dependent upon our computers. I, for instance, could no longer write books like this one without the facilities offered by a word processor. In much the same way that cheap electronic calculators did away with our children's need to learn their multiplication tables, the vast amounts of computer storage available allow us to hold our less important memories in this form rather than in our heads. Ultimately, maybe within just a few decades, the majority of our memories will be held in this way; and we will no longer need to maintain diaries as source material for our autobiographies. Indeed, when chips are implanted in our brains, these remote memories may be able to hold the records of the inputs from all our senses and emotions. It will then be possible to call up whatever event from the past we want to experience – our first love, say – and relive it exactly as it happened!

This poses a number of philosophical and ethical challenges. Where our identity is split between our physical being and our electronic *alter ego*, who is to say who, or what, we are? It will take some time before this new *Homo integrans* is accepted: psychologically by the individuals involved, and legally by society. At the other end of the spectrum of potential problems, there will be those who, for a range of macabre reasons, choose literally to live in their memories. There will be those who go even further, to live in other people's memories. All of this offers enormous potential which was also the case when our ancestors managed, for the first time, to share experiences through the new invention of language. That is why we see it as a new stage in evolution.

Taking the opposite tack, looking perhaps for computers rather than ourselves to become the next stage of evolution, there have been a number of workable Artificial Intelligence (AI) systems developed for practical use, in particular in the areas of medicine, oil exploration and computer control – albeit that all of these have been 'expert systems' rather than full-blown AI. There have also been experimental 'neural networks' of connected microprocessors. Based on the practical difficulties in implementation I have observed with relatively simple applications, I suspect the problem of learning is often more complex than is allowed for. Important elements may emerge earlier to handle the human/computer interface – as it is already the case with 'help-desk' operations; and another three decades of development in IT might solve all these problems!

Robots

One area which is frequently chosen as a particular feature of future society is the emergence of robots into general use. Marvin Cetron (1994), for instance, follows one popular theme when he predicts that 'personal robots will appear in the home of 2020'. Pearson and Cochrane suggest that 'domestic robots will be small, specialized and attractive, for example cuddly.' Cuddly!

Robots have, indeed, already proved very productive in certain production environments, especially in hostile ones, but they have not yet diffused as rapidly into other environments as the pundits have predicted, and there is little evidence that they will in the near future. IBM found that human labour was more productive even for assembling PCs, and threw out a billion-dollar investment in robots.

Possibly the problem is that of 'anthropomorphism' ; it is understandable that we would want to create our technological children in much the same image. Yet the most successful 'robots' to date have no physical similarity to humans. One only has to think of the domestic washing machine, which has replaced the housewife at the sink; or the auto-pilot, which now even lands planes in conditions which would be impossible for a human to consider. Thus, the most important aspect of robotics may well be that of adding intelligence to the machines which already serve us.

As the major driver which has already emerged, Information Technology will have an increasingly revolutionary impact, but the new communications technologies will have an even greater impact.

The one remaining area of hardware where significant development should be expected is in the human/machine interface; by the second decade of the twenty-first century, this is likely to go far beyond virtual reality to offer quasi-symbiotic links which will dramatically extend our powers not just those of the computer.

Computerized aids to human work will continue to increase in number, probably quite rapidly, but few of these will look anything like the popular conception of a robot.

The Software Jungle

Turning to the problems which may hold back the IT Revolution, the major area where developments are rather more questionable is that of software. John Taylor, of Hewlett Packard, makes the point that the amount of software in the field multiplies every decade by something between 1,000 and 100,000 times (and the amount of information held on databases by the same amount), something like ten to one hundred times greater than that for hardware. Thus, there will be quantities of new software, but of what quality is less certain. Yet this software must develop effectively for all that hardware power to make sense.

What we see is what the programming – human or machine-produced – offers us at the interface; the hardware is hidden. It is the development of the software which will really deliver the IT

Revolution. So far the recent efforts have been puny, at least as far as the average consumer is concerned. Thirty years ago computers were, in effect, putting a man on the moon. Since then they have only been used to put a combined typewriter and calculator on our desks – now, at last, also combined with a 'data telephone'. Yet even these simple uses have often posed major problems, as anyone who has had their machine crash at an inopportune time will be able to testify.

The big breakthrough will come when software can deliver what we as individuals *need*, not just what the software vendors find easy and profitable to make. Some more sophisticated offerings are already becoming available. The importance of the World Wide Web (WWW) is now almost universally recognized. The big advantage of WWW is hyperlinking. My own offering on the Web, for instance, effortlessly switches users between pages on our own mainframe and those of our partners around the world. We are now planning to integrate all our systems in this way. Users will be switched seamlessly between the Web, email, CD-ROM and a variety of other media without even realizing that this is happening. The computer is becoming an invisible part of this universal access to information – exactly as it should be.

The technical problems come on two fronts. One is the sheer complexity of programming such personalization implies. If you have ever tried to use a sophisticated package, you will realize how confusing the resulting myriad of features can be. It can literally take days to find what you want – ploughing through hundreds of pages of badly written manuals – and so you too often settle for something less. What is needed is, as we have seen, some form of artificial intelligence.

At the other extreme, it is true to say that, as yet, the vast majority of programs are relatively crude. Further, even in relatively simple systems, programmers make errors which cannot always be predicted. As the complexity of programs increases, so may their instability – with awesome consequences, as errors destroy organizations and lives. The 'Millennium Bug', caused by early programmers reducing program sizes by limiting the space allowed for date (but in the process making them crash as they moved on to the year 2000) is just one well-known example which has caused billions of dollars' worth of problems. 'Safety' testing of programs may, therefore, become a major industry in its own right. Beyond mere stability, security has become a preoccupation with most system designers. Computer crime, that which is detected, has been increasing as rapidly as the spread of computer communications.

The final caveat is about 'languages'. There is, as yet, no one language which we all know, or need to know, to accomplish what we need on all the packages we own. Microsoft is trying to set some standards; but, as even these standards are heavily copyrighted, one suspects the intentions may not be wholly altruistic. Its competitors are, similarly, trying to use JAVA as a means of undermining its position. We are in the position of many small tribes who share no common language, but who must learn a new language every time they travel more than a few miles. It is reasonable to expect that a common language set will emerge or will be imposed by regulation.

Much of the new software will be in the form of small-scale offerings produced by small, independent suppliers to meet the specialized needs of small groups of consumers. This, along with the need for many small information providers, will change the shape of much of industry. On the other hand, the core offerings, which these suppliers will use to build their own offerings and which the consumers will use to interface with these, will almost certainly be provided by the large corporations – Microsoft and a revamped IBM, for instance – since only they will be able to afford the scale of investment needed.

The scale of investment may, in effect, reward some of these packages with a form of monopoly. This will be addressed in two ways. In the first, the standardized interfaces – changes to which will need to be agreed, and publicized, well in advance – will allow competitive suppliers the chance to reverse engineer their own offerings to match those of the monopolists. Where this fails, the second answer will be – in view of the global importance of these packages –

that they will be subject to regulation. Microsoft is already attracting as many lawsuits as IBM did in its heyday. No doubt everyone, or at least the handful which remain in the mass market, will still be allowed to earn handsome profits – but no longer obscene ones!

The main driver, or here possibly the main limiter, will be that of the development of suitable, and *safe*, software – probably using artificial intelligence to handle the interfaces – which will extend people's capabilities. Modular languages may be a first step in this direction.

Although the emergence of many small suppliers will change the nature of industry, the core packages will be produced by the large corporations, who can afford the investment – though their monopoly position will be held in check by standardization and regulation.

The Communications Revolution

As already stated, the most revolutionary development in technology is now represented by the exponential growth in our abilities to communicate. Later we will look at how such processes may change not just the way we work but the whole structures of organizations. What is perhaps even more important, however, is the way in which they may affect our own lifestyles – at one extreme – and the workings of the whole world at the other. Thus, one outcome of the Communications Revolution, which may develop more slowly than some predict, will be the shrinking of the world. At the level of the IT users (and their number is likely to grow to a billion or more over the next several decades), there will be no boundaries of any type, let alone national boundaries, in their world. The global village, much promised since the 1960s, may at long last make its appearance in practice as well as theory.

Almost all our groups emphasized computer communications as distinct from IT. Almost all mentioned that communications will be totally mobile (by 2015); with comments such as 'mobile phones include a miniature video screen'. One group, though, listed a very interesting concept: 'telecomms are free!' This is one of the 'wild cards' and may sound silly in an age when telecom multinationals are among the most profitable commercial enterprises. On the other hand, it starts to make a lot more sense when you realize that the data they carry will be the lifeblood of a future society, as water is an essential commodity now; and, like water, once the distribution infrastructure is paid for, everything else is effectively free. Thus, such communications really are likely to become so cheap as to be effectively free. Even if the carriers do not recognize the logic of their position, in their attempts to stimulate technological and economic development some governments will change the pricing structures: a fixed connection charge – where connecting even fibre optic cables to the home is cheaper than connecting water supplies – and everything thereafter almost free. Once one government, or even one carrier, offers this competitive advantage, others will follow.

This breaking down of IT boundaries will eventually carry across to the cultural and physical ones; it will seem nonsensical if you have to obtain a visa to visit physically someone to whom you talk electronically on a daily basis. In any case, satellite television has already made it virtually impossible for any government, even the most draconian of military dictatorships, to seal its borders. This will shape the world, hopefully uniting it, where national media have often been used to divide it. It is difficult, though not impossible, to hate those you work alongside – even if in cyberspace – and certainly it is much more difficult to go to war against them.

Richard Tutner (1994) writing in *The Wall Street Journal* imaginatively, and enthusiastically, suggests that: 'Consumers, for instance, could have instant access to every movie and TV show and piece of music ever produced. They could get everything in the US Library of Congress. They could go shopping with a sister in San Francisco, examining products in a "virtual mall" from every angle. They could compare hotel rooms for a vacation. And they could exchange information with people anywhere in the world, about narrow topics they thought only they were interested in. And that's just the beginning!' From the range of offerings he promises, his description does give a good feel for the potential this particular aspect of the future might offer.

Perhaps a more typical example of the most productive use of the Internet is given by Gary Stix (1994) when he describes its use by physicists. 'The computer ... has become, in effect, a daily wire service for high-energy physics theorists as well as researchers from more than 10 other disciplines, primarily in the physical sciences and mathematics. Every day 20,000 or so electronic-mail messages carry the abstracts of new papers stored in the computer's databases to more than 60 countries. Readers of the summaries then download thousands of copies of the full papers.' This gives a real indication of the power already available. Previously, such papers would have taken a couple of years, not just a couple of days, to reach those in the field who needed to read them.

Returning closer to the short-term reality, however, not merely will it take some time – probably several decades – to reach the ultimate levels of sophistication promised by Richard Tutner, but there are really two distinct, and very different, concepts involved in the current technology.

The super-highway, promoted by the media owners (especially those in television, studios and cable channels), is characterized by a large bandwidth capable of handling moving (cinema-quality) pictures: centralized, commercial provision, for a fee, of one-way entertainment – 1,000-channel TV and movies on demand. The Internet, on the other hand, is a low-bandwidth, text based, networked, co-operative, free of charge interactive communications system – with tens of millions talking to each other.

The short-term problem is that these two systems, the super-highway and the Internet, could not be more different from each other. In fact, the two approaches need to be combined to offer what is really needed. As indicated above, the full bandwidth of the (fibre-optic) super-highway will be essential to handle the massive information flows to be expected (not least when many of these will be in the form of moving pictures, even if of videophone format rather than movies) and the interactive nature of the Internet will be needed to make the whole thing work in the way that individual users will demand. The form will I suspect be a hybrid: local networks supported by high-bandwidth backbones; and this is what is now actually happening on the ground, but there will probably be no such compromise on charges. The services will ultimately have to be paid for. Remember, though, the earlier comments about virtually free communications, and this becomes much less of a problem. Thus, the level of charges for even the most sophisticated use may well be within the comfortable range of most potential users.

John Heilemann (1994) reports that 'Critics predict an information glut. A television system of unlimited channels, they say, will flood people with more of what is already drowning them. Others are less afraid of immersion than seduction. No one who has played video games or used the Internet doubts their addictiveness. In the past, television was the opiate of the masses, goes the argument; in future it will be their crack cocaine.' Intriguingly, Heilemann adds something missed by most other commentators: 'The next step will be assembling new programmes on demand – a montage of Humphrey Bogart's best movies or news reports of fires in Los Angeles.' This may be some time off, though maybe not so far away if the assemblers are human and located in downtown Bangalore, say, and will add an extra dimension to the whole concept of home entertainment. In any case, the investment is already so high, and the need so obvious (indeed, a social necessity), that the problems will be sorted out; the super-highway/Internet will not be allowed to lapse. Once started, nothing can stop it!

A particular aspect of communication is that of the more sophisticated forms of teleconferencing which allow meetings to be personally 'attended' by participants from anywhere in the world. Ultimately perhaps, the use of large (3D) flat screens – as predicted by Pearson and Cochrane (1995) – may also convey the impression that you are sharing the same place with the person at the other end of the line, be it a 'meeting' or an individual videophone call. This will bring you much of the benefit of an actual face-to-face contact; and, where groups (of family members or friends) are now so dispersed, this may become an especially important device for maintaining social contact in a fragmenting world.

Global communications will, at long last, create the global village – among IT users – and this should open all borders and allow for better understanding between peoples and nations. Remote video contacts, teleconferencing or individual videophones, are already a reality.

There is considerable confusion between the promoters of the super-highway and those of Internet, which in (technical) fact should not be seen as similar concepts, let alone as compatible systems. Even so, the rewards are so high that the problems will be sorted out, probably by offering a hybrid combination of the two, to form the backbone of the information society.

Organizational Structures

Perhaps the most immediate impact will, though, be on the structures of organizations – in the West at least. Information, and control signals, have traditionally flowed vertically in such organizations. In the past we all have had our bosses, and some of us have had our sets of subordinates too. The new communication paths already are horizontal; this is the way that such IT-based communications flow and there is no way that any manager can halt this! In future we may just have our peers. Networking in this way has to be clearly recognized as a new form of communication, with new rules.

One of the effects of this new form of communication has been the redefinition of many managers' roles. Many have lost their jobs since a significant part of middle managers' roles was to provide information to those below them in the hierarchy and, with that information now provided direct, this role no longer exists. Indeed, increases in unemployment are often blamed on the IT Revolution. It is a convenient scapegoat; 'structural unemployment' is supposed to reflect an underlying shift in employment away from those who do not have the requisite IT skills.

In the specific context of 'structural unemployment', *The Econ-*

omist (11 February 1995) points out that 'In the past 200 years millions of workers have been replaced by machines. Over the same time period, the number of jobs has grown almost continuously, as have the real incomes of most people in the industrial world. Furthermore, this growth and enrichment have come about not in spite of technological change but because of it.' It goes on to provide more recent evidence. 'Despite a huge investment in computing and so on over the past decade, unemployment in the United States, at around 5.5%, is currently no higher than it was in the 1960s . . . This is hardly a persuasive sign that IT is a big cause of unemployment . . . From an individual's point of view, process innovation (making things more effectively) . . . may indeed reduce employment. But the economy as a whole will enjoy compensating effects.' It seems, to us at least, that the current unemployment problems can just as easily be related to the 'revolutionary pains' mentioned in the opening chapter. Once we have fully entered the new order, and stability has returned, full employment may also follow.

More important, though, is the impact it has in general, even on the workers at the lower levels. The effect here is to give them much more flexibility and power. The best analogy is with Japanese structures, where such horizontal communications have long dominated.

The Communications Revolution is already having a dramatic impact on organizational structures, leading to a redefinition of managers' roles and a move to softer HRS (Human Resource Strategies) in terms of relations with staff in general. This will subsequently extend to social structures in general.

Publishing

The industry sector which will probably experience the most changes as a result of the IT and Communications Revolutions is that of publishing. There will inevitably be a dramatic leap in the volume of material published in one form or another. The most important change, though, will be in the form; printed publications may still expand somewhat in volume, but the exponential growth will be found in the electronic forms of publishing. It is now quite possible to imagine a book-sized (and book-weight) computer (something like a notebook computer) which will show you (in the same detail and with the same clarity as a book) a whole page of print. It will be as easy to read this as a normal book, even easier in poor light. It will, however, also be possible instantaneously to look up references through cell-phone/modem connections. Of course, the 'book' could, at the same time, allow the reader to keep in touch with any messages; and an ear-piece (which would in any case be needed for multi-media material – another advantage) would allow it also to operate as a mobile telephone (possibly eventually even a video-phone). Even so, it has to be admitted it will still be, in its most important respects, a book! For some purposes a picture may be worth a thousand words, a phrase often on the lips of multi-media suppliers, but for many others – especially in the worlds of ideas – a sentence may be worth a thousand pictures.

The form(s) of publishing will probably cover a wide spectrum, far more so than at present. It may offer, seamlessly in one package: material displayed on a screen – words, pictures and diagrams – probably animated in key areas; video – possibly in 3D; and audio. The exact form will depend entirely upon the requirements of the publisher and the authors, which may become a team. The biggest beneficiaries will be the consumers. We will have the whole (electronically published) world at our fingertips, in whatever form we find most useful.

This will make knowledge in general available to a much wider audience and in much greater depth. Ian Pearson (1997) nicely encapsulates these developments by making the point that, where the first Industrial Revolution removed the need for physical labour by humans – and with it the advantage enjoyed by the strongest men – the current revolution is removing the need for the knowledge held in human heads, removing the advantage currently enjoyed by our most knowledgeable citizens. What skills will be needed next is less clear. Possibly these will be communications skills or possibly integrators or synthesizers, who can take disparate pieces of information and (with a soupçon of creativity) turn them, synthesize them, into something more valuable.

Perhaps the most revolutionary form of the new publishing will be that undertaken on the small scale. It will be possible for potential authors to produce their 'books' (then, of course, more likely to be as a multi-media package) themselves, using the resources (of information acquisition and manipulation) described earlier. Then they will be able to offer them for sale through the networks.

The publishing industry will possibly see the greatest changes of all, with massively expanding sales volumes, based on new, electronic forms of presentation. This will bring in major new providers but, especially, will dramatically improve the availability of published information for consumers, opening it up to more of them, in greater depth. The most revolutionary aspect may be the emergence of individual publishing where the author will also be publisher, many publications selling just a few hundred copies or fewer.

Technological Highlights

As we end the chapters on technology, it is worth repeating that the key fact in this scenario is that almost all its resources are now in effectively unlimited supply. The result is a very optimistic view, with a nicely incremental growth in the many new things we will soon find ourselves able to do. Ultimately we will be able to do whatever we choose to do. Not least, medical advances will significantly extend our lifespan – perhaps beyond 100 years – so that we will also have the time to do what we want.

The exploration of space – and then of its colonization – will bring massive impacts in terms of the availability of physical resources, and even the development of society in general. Not least, it will give us new – physical – frontiers to be expanded for the benefit of humankind.

The ability instantaneously to access – no matter where you are – all the information stored on databanks around the world, and to communicate with all its inhabitants, will radically change individual lifestyles and the structures of society as a whole. As the resource – communication of existing information – is almost free, once one has the necessary electronic links (which are even now cheap enough for a large proportion of the world's population to afford), the impact of this revolution may reach its peak, worldwide, much earlier than many expect – perhaps in just a decade or so.

Not least, the symbiosis of humans with computer networks – with the existing slow interfaces through the existing terminals or much faster ones though the brain–computer links yet to come – will, in effect, produce a new level of human evolution: *Homo integrans*.

5

SOCIETY'S WINNERS

These days it is very easy for us to talk about the technological changes occurring around us. We have become nonchalant, even complacent, about these changes and would feel something was wrong if they did not regularly feature in the news media. Having initiated the massive changes which are under way, these technological marvels no longer represent the main drivers for the changes to come. Now it is the structures of society which, above all, are being torn apart around us to be replaced by very different frameworks. We feel much less comfortable about these changes, for they represent the true frontiers of change.

As we will see, the most obvious signs of these changes are not the increased levels of material greed promoted by politicians of the right but the intangible freedom, to decide our own futures, now being grasped by individuals. Across much of the developed world these new freedoms are, for the first time, being made available to most of the population. Above all, the largest 'minority', that of women, will not merely achieve equality but will attain some degree of advantage as society overall is progressively feminized. We are entering upon what may come to be described as the 'women's century'.

The changes described in this second part of the book are, in many ways, almost diametrically opposed to those in the first part.

The most obvious of these contrasts is that between the hard realities of technology and the soft values of the individual, but there are others. Thus, those in the first part are largely incremental in nature and in many respects quite predictable in terms of their impact, whereas those in this part are likely to progress from one major discontinuity to the next, leading to considerable uncertainty. Indeed, as we will see, this is the part of the book which contains the largest number of unanswered questions. This is one reason why so many people are worried about the future, about the new threats which seem to be hanging over us; the unknown always seems much more menacing. Even so, though the detailed outcomes are unknown, they are likely to be optimistic ones since the general message of this part is the empowerment of the individual.

Individualism and the New Role of the Community

Indeed, the major social force is now encapsulated in a powerful idea – empowerment of the individual. Some 70 per cent of our individuals expected this to be a dominant force in society within twenty years.

Some might say that this development first started, in the 1980s, as part of the search for personal affluence. On the other hand we, and the managers in our research groups, believe that such greed will not be the most important aspect in the longer term. Almost two-thirds of individuals, for instance, believe that materialism will be replaced by ethics within twenty-five years, and three-quarters think that organizations in general will stress ethical values within twenty years. Direct evidence for such a shift in mood came not from the political swings in this direction to be seen around the world but from the members of the largest building society (savings and loan institution) in the United Kingdom, who in mid-1997 voted to remain non-profitmaking despite the prospect of receiving an immediate cash handout (bribe?) of more than £2,000 each! Thus, we believe that the emergence of what we would call true individualism – a search for personal fulfilment largely untainted by selfish avarice – is the most important social trend at work.

This may be seen, for example, in a new definition of equality: the right, the freedom, to be *different*, not the demand to be the same. Almost two-thirds of individuals expect there to be a universal bill of rights by 2030. This is a new development, where there had formerly been a tension between the rights of freedom and of equality. Previously, an improvement in one usually led to a deterioration in the other. Now, with the benefit of the IT Revolution, which allows us to deal effectively with individuals, not just the masses, freedom and equality can improve simultaneously. The new battle-cry will, therefore, be seen in terms of the individual's basic right to fulfil his or her full potential and aspirations.

It is often claimed that such individualism is inevitably at the expense of group cohesiveness, especially that which holds communities together. It is true that the individual is no longer subservient to the will of the group, but this is a fundamental misconception of the new developments. After the current phase of uncertainty and confusion about such relationships, a new relationship between the individual and the group will emerge. Three-quarters of individuals expect new forms of community and values to emerge by 2025, the same result as is recorded for personal values. It is not yet clear exactly what form this will take, but it seems likely that it will be based more on peer-to-peer relationships rather than the hierarchies which typified the previous regimes. It is also likely that this will be a richer relationship, a pluralistic one which is allowed to take the form best suited to the specific situation. Over the longer term, full flowering of this individualism will represent nothing less than a quantum leap in social organization. In summary, the philosophy of individual empowerment is likely to be the greatest force for change in the next Millennium.

New Lifestyles – Post-Materialism, Post-Fordism, Post-Modernism, Post-Everything?

One possible starting point for exploring the subject of individualism is the academic debate about post-modernism. The immediate stumbling block is that the term is used by different people to mean so many different things, usually with very specific connotations and too often in impenetrable academic jargon. Indeed, perhaps the real problem is that we are only at the beginning of the revolution, so it is very unclear exactly what post-modernism might ultimately become.

At the heart of modernism was the depersonalizing move from the country into the city, to work in the new factories which were the powerhouse of the first industrial revolution. With this move gradually came secularization, a Catholic God replaced for many by the Protestant work ethic, which had to be demanded of those working in the increasingly machine-driven environment of those factories. Many of these developments are reflected in the dominant social science of that age – economics (in which I was trained).

Post-modernism, therefore, represents the breakdown of this system of values. The factory system required that the workers followed their instructions exactly and physical labour, along with the discipline needed to ensure effective organization of its application, was the essence. The processes of that era, which in too many organizations still hold sway, had the worker service the machine so that this might be productive. In the knowledge society, intellectual effort accompanied by creative interpretation of the rules is becoming the norm.

The process of social change can, perhaps, best be seen in the workplace, where the change can be summed up by the move to the Human Resource Strategies (HRS) we looked at in the previous chapter. When we are moving from a surplus of workers to one of a scarcity of skills (in the West at least), this is most clearly encapsulated in the recognition that the prime investment is now made in people rather than in machines; and the whole production process, in offices as much as in factories, has increasingly focused on making individual workers more productive. This has reached its latest peak in terms of the increasing debate about 'intellectual capital' – the value of the intangible knowledge assets held by the organization, typically in the heads of employees.

Beyond the workplace, relaxation of the more excessive demands of the industrial system has been reflected in changing individual lifestyles. Some sociologists see the most important aspect of post-modernism to be a move from a society based upon the acquisition of physical goods to one which places a greater emphasis on *intangibles.*

The role of the group, while declining in social power overall, is still inherent in terms of its supposed cultural control over the behaviour of the individual. At work, employees – increasingly now knowledge workers – might determine their own, informal contract with the organization, but this is still typically placed in a group context. At home, their geographic neighbourhood, or class group, may no longer determine their behaviour, but they commit instead to new groups based on interests such as football, or single issues such as the environment.

The essence of post-modernism seems to be the developing freedom of the individual to create his or her own lifestyle(s) without regard to the groups which would traditionally have exerted (cultural) control over this; in particular, individuals are rapidly moving outside the control of the industrial system.

Lifelong Education

The most important formally organized force for change in personal circumstances, and hence for genuine empowerment of the individual, is that of education. More than half our research groups stressed its importance. Indeed, it can be argued that the exponential growth of education, around the world, has often created the social revolutions now taking place. Not just the parochial view of those in our research groups, it is the view held by most governments across the world. The problem is, as we will see, governments largely do nothing to follow this belief!

To date, almost all increases in education have mainly taken the form of ever-higher levels of basic education being delivered to an ever-greater proportion of the population. The Far Eastern Tigers and, especially, Japan – which has put its money where its mouth is – have pursued this process further than other nations and have reaped the benefits.

The difference is not between those nations which educate their populations but between the levels to which they educate them; between the leaders, in the Far East, and their more conservative erstwhile competitors in the West such as the US and UK which, largely for reasons of market ideology, are less committed to the human infrastructure. Even these laggards, however, are increasing their investment in basic education. To put this in context, Peter Drucker (1993) documents the enormous scale of this investment.

> Formal schooling – schooling of young people before they enter the workforce – takes about one-tenth of GNP (up from 2% or less at the time of World War I). Employing organizations spend another 5% of GNP on the continuing education of their employees; it may be more. Around 3–5% of GNP are spent on research and development, that is, on the production of new knowledge. Very few countries set aside a similar portion [up to 20%] of their GNP to form traditional [that is, money] capital.

As already indicated, the difference between nations is the rate at which such investment is being made. A high rate of investment is needed to support a high rate of change within an economy. The level of investment can now be significantly reduced, by the use of new technologies. Using these, for example, we, – at the Open University (OU) – are now able to teach hundreds of undergraduates, even in the developing nations, who otherwise would not be able to afford conventional education. When discussing the topic in more detail, almost half the groups saw such distance learning –

the technique where the OU leads the world – having a major role to play. It may be, therefore, that in future the most likely location for mass higher education, especially that concerned with ongoing education, will be in the home (or in the office) rather than in a college, and 80 per cent of individuals foresaw this happening within twenty years.

One particular reason for the dramatic changes in location and in teaching styles arises from a less obvious educational requirement. This is the impact of ongoing education. Most educational systems are still happy to see their responsibilities end once the individual is accepted into the workforce.

The emerging requirements of the knowledge society demand rather different solutions, however. Thus, the rate of change is now so great that an individual will have to take on a number of 'trades' over his or her lifetime and unless they are regularly retained (indeed almost continually re-educated) to take on each new trade, they will become almost unemployable. Rosabeth Moss Kanter (1994), one of the leading gurus at the Harvard Business School, makes the point that individuals will no longer have the benefit of lifetime employment. 'If security no longer comes from being employed, it must come from being employable. . . . Continuing upgrading of skills and pursuit of opportunities is a lifelong proposition even inside a single corporation.'

The truth of this predicament has now become obvious to the leaders of most nations. Unfortunately, the dilemma this poses is: who pays? Should not the employer or the employee pay; for will they not be the main beneficiaries? In Japan, where both of these parties already have a sound understanding of what is needed to create a better future for all, this poses no problems; and the re-education is happily undertaken as part of the normal working environment. In the West, however, few employers – and even fewer employees – see the point of such retraining or re-education; and almost none of them are willing to pay for it. Like it or not, governments will usually have to take the first step. Those nations which do not address the problem will rapidly fall behind their international competitors; on-going education will probably become the largest single industry across the globe.

Education is already almost universally recognized as the most important investment any nation makes; the difference lies only in the degree to which individual nations match the theory with hard cash. Basic education will soon extend – worldwide – to ongoing, lifelong education in order to meet the demands of the various economies, forced by very rapid change. Such levels of ongoing education are increasingly becoming an important *right* to which individuals are entitled.

New Work Roles

What we have not yet looked at, in detail, is how the organization of work has already changed. This is even more dramatic than the move to investment in the individual rather than in the machine would imply. 'Fordist' assembly-line work was determined by exact rules, encapsulated in the Taylorist theories of scientific management and, especially, in the discipline of work study. Here, the optimum action, for every step of the production process, was determined; and the worker was required to repeat those few steps over and over again, possibly for years on end. It was clearly soul-destroying. On the other hand, combined with an emphasis on strictly hierarchical structures of management, it enabled relatively unskilled, untrained workers to carry out almost every task needed.

The new knowledge workers – rapidly growing in number, to become the great majority of the overall workforce – are required to handle such a variety of different tasks that it is becoming impossible to set similar rules for these. Even the few workers who remain on production lines now have to be multi-skilled, so they can switch from one task to another at a moment's notice. From the point of view of the workforce, especially those who are unskilled or semi-skilled, the change has been dramatic. The demand now is for fully skilled, and even for fully educated, workers. If you visit the Toyota assembly lines near Nagoya, in Japan, you will find few assembly-line workers who have not completed at least fifteen years of education, sufficient to qualify them for a university place (perhaps even a degree) in the West. Even after that, they have then been continuously trained by Toyota itself.

The downside is, of course, that the unskilled and semi-skilled are fast becoming not just unemployed but *unemployable.* Since 1980, in the US at least, the demand for labour has only increased among the top 10 per cent of the most skilled workers; for all other sectors demand has dropped off. In the West, where far less attempt has been made than in Japan to extend education to this group, this is seen as an impossible problem dignified by the term 'structural unemployment'. This implies that it is an unavoidable sign of the times, which as Japan shows it is not, rather than a failure by the leaders of society, in particular its politicians.

One result of this multi-skilling process has been a dramatic increase in the degree of self-management. It has also resulted in a greater level of team-working, largely based on peer-to-peer relationships, with the manager now working as a team leader. The ultimate development of such self-management, now being implemented for instance by consultancies (which, one way or another, are accounting for a growing proportion of the workforce), is the emergence of the (academic) collegiate form of organization. In such a system, within limits, we are free to manage all aspects of our work life ourselves.

A less obvious trend is that the general nature of each individual's set of tasks will change over time. The definition of the individual's role, and that of the team within which he or she operates, is therefore becoming increasingly flexible and variable. No set of rules, nor any hierarchical management structure, can cope with this. One informal outcome is the cellular organic structure. In this form of lower-level management structure, the relationship between members of the team and between the team and other parts of the organization is constantly shifting, being renegotiated almost daily. Even the boundaries of the team, of the 'cell', change – taking in other members of the organization, for example – as the circumstances change, in much the same way that other living organisms do. In this context, no less than 85 per cent of our individuals expected to have 'multiple careers' within twenty years, and this was reflected – in a more specific context – by a similar number who expected contract working to be general, a much more radical departure.

Traditional forms of rules are clearly unworkable in such situations, so control is exerted (often by default) by the culture of the organization, and of the parts of it closest to the 'cell' (team), in much the same way that DNA governs what happens to the living cell. A more general form of this wider set of linkages has often been referred to, in recent years, as 'networking'. Such culture-driven organizations are already particularly evident in the high-technology sectors, but the form is spreading to more and more organizations as the rapid rate of change demands ever more flexible management.

In changing times, those workers who are not trained to met the changed needs will be unemployed, and unemployable. This is not an inevitable process; it merely requires that suitable training be provided; and it is likely that the rest of the world will eventually follow the lead of Japan – and unemployment will slowly reduce to the levels seen in the 1950s and 1960s.

The growing need for personnel to be multi-skilled, and often to adopt creative approaches to their work, means that self-management will increase rapidly. Indeed, the new knowledge workers will have to accept an increasing degree of self-management in order to achieve the flexibility demanded by the variable tasks they encounter and organizational structures will need to change to reflect this, ultimately in some cases – especially in the rapidly growing fields of consultancy – moving to peer-to-peer collegiate relationships. To cope with the flexibility needed in the workplace, cellular organic structures – which continuously adapt their internal and external relationships, led only by the culture which envelops them – may be increasingly found.

Group Breakdown

Although team-working will remain, and may actually grow in importance, the flexibility implied by the new organizational forms means that long-term group identities will be weakened. The individuals may appear to be just as intimately involved with those around them, from day to day, but their fellow team members may be different people from those they worked with a year ago, or perhaps even just a week ago. Group solidarity, in these new circumstances, will change its nature.

This represents an important development since, for the past century, groups in one form or another have been the building block of society. Individuals, in this societal context, have been largely seen in terms of their group membership, and typically saw themselves in these terms. Indeed, the group to which people belonged, and whose behaviour they were culturally bound to follow, was often the same at work as at home. Thus, in earlier decades, the factory was often surrounded by the houses where its workers lived as a tightly knit community. Son followed father into the same job.

This represented in many respects a continuation of rural society, where relationships with neighbours accounted for almost all the contacts made by an individual and lasted a lifetime. In this environment, the linchpin of society was the extended family; brothers, sisters, aunts, uncles, cousins – and often the neighbours as well – all sharing a very strong common identity.

Nuclear Breakdown

In the years following World War II, however, the structure seemed to change dramatically with the widespread emergence of the nuclear family. As people moved, and were moved – not least in ambitious slum clearance schemes – the old communities fragmented. These moves were not just the result of changing work patterns, though the increasing mobility required of labour played a part, but were also the outcome of the consumer society. This demanded that families, now father, mother and just two children, moved from the less desirable accommodation inhabited by their parents to better homes. They also developed more sophisticated aspirations, which added social distance to that of geography. With increasing affluence, and better education, families became smaller, so they were inevitably less extended than before and the nuclear family became the new ideal.

Now the 'model' is changing yet again. Helen Wilkinson (1994), reporting results from the Demos 'Seven Million Project' which is researching the 18–34-year-old population (men and women) in the United Kingdom, says that: '. . . there has been a long-run swing away from traditional concerns for security, authority, rigid moral codes and a belief in the centrality of the family, that dominated the value map a generation ago.' Even the seemingly sacrosanct nuclear family is splintering. Helen Wilkinson again reported (in 1994) that: 'Over half today's 25-year-olds have cohabited with a partner compared with 1–2% twenty-five years ago.'

What is rarely discussed, however, is the extent to which the nuclear family itself was an artificial construct. Crucially, it depended upon the idea that the only role of the wife in this family was that of home-maker and mother, looking after the needs of the (two) children and husband. Where the previous role of women had primarily been to work, often alongside their husbands – with the children looked after by other members of the extended family (especially the grandmother in retirement) – the new affluence allowed the luxury of the wife staying at home. There, she took on the role of the servants, who had by then become unaffordable for middle-class families, or emulated it in the lower classes.

The new role started to become dispensable when, after the 1950s, home appliances increased their penetration of households. At the same time, the growing pressures of the consumer society meant that most families found they needed ever-higher levels of income, which could only come from a second earner. Paralleling this was the growth of the service sector, with its urgent need for part-time female labour, and an offer emerged which few women could resist.

The core structure of the nuclear family, with the wife as dedicated home-maker, had thus disappeared long before commentators began to notice any disintegration and less than a generation after the model had become accepted as the natural order of things. In any case, the idea of a perfect match, interest-by-interest and lifestyle-by-lifestyle, between two individuals becomes less and less likely as they progress through the years of 'marriage'. As this can now add up to more than half a century, and they will each mature in different ways, perhaps the reality never lived up to the myth. Perhaps it was just a blip on the long history of the extended family as it made a natural transition to a different form. The newly emerging 'extended' version is now just as likely to comprise the families inherited from previous marriages, step-fathers and step-mothers rather than aunts and uncles. Perhaps this emerging new form of the extended family will ultimately provide the same level of support as the more traditional version. The breakdown of the nuclear family has significant consequences for society in general, as well as for the individual. If nothing else, such a breakdown undermines many of the myths which recently have come to underpin Western society.

The changing role of the housewife may be central to this breakdown. Even so, it is not clear, despite the benefit of hindsight, whether feminism drove some of the changes leading to the breakdown of the family. Maybe it merely occurred at the same time, driven by much the same forces, and I tend to favour this view. Whatever the reason, the drive for gender equality – across all parts of society – irretrievably changed the position of women in the West, and by example elsewhere as well.

Over the past century, there has been a seemingly natural progression from community to extended family to nuclear family to individual. Whatever the explanation, the trend, which dooms the nuclear model, has been very clear; and there is no reason to think it will not continue, though it is possible that new forms of the extended family will emerge.

As it was as much myth as reality, the lamented demise of the nuclear family and of its values is possibly less important than at first sight it might appear.

The New Power of Women

Perhaps one of the most distinctive features of the twenty-first century may, in general, be the growing power of women. Two-thirds of our individuals expect gender equality to be global by 2030 and, more importantly perhaps, a surprisingly high proportion (70%) expected a 'notable rise in feminine values' by 2015. Indeed, I would argue that the most successful new political movement of the last half-century has been that of the feminists. Their impact upon the Western world has been much greater than is appreciated, even by the most earnest of their bra-burners. In many respects they, almost alone, are now setting the truly revolutionary agenda for the future. Helen Wilkinson (1994) points out that: 'Women have been at the forefront of changes [in society]: their values have changed most and have come to define the overall shape of British values.' She continues, 'As a group their values are becoming androgynous: they are much more willing than men to think flexibly about gender roles.' On the other hand it is fair to say that these views, no matter how well grounded in research results, are not yet universally accepted.

In view of their historically weak position, in terms of their power to influence the future of society, the dramatic nature of the rise will largely reflect the very low base from which it is starting. The real shifts in power resulting from the coming changes will have significant impacts on the development of society as a whole, not just women's position within it. Warren Wagar (1992), for instance, explains one key aspect. 'Their characteristic leadership style focuses on acquiring and wielding power for the good of the organization rather than for purely personal gain. It will give pacesetting women the edge they need to displace men as the dominant force in the economy and in government.'

It is difficult to predict the final outcome. It is, though, quite easy to suggest that most women will at last start to fulfil their real potential, and that this potential will be defined by them, not by men. It will probably take several generations before women can fully escape from the cultural trap which society has set for them. For centuries women have been shaped from childhood by society; it is seen as only natural that they will want to play with dolls, preparing them to be the home-maker. The result is that even women who have managed to break free from the adult stereotype sometimes feel guilty that they are not fulfilling their natural role as mothers. What should really be seen as their childhood, and adult, models unfortunately is not yet obvious. It clearly is not to become surrogate men. Maybe it will turn out to be not too different from the softer feminine virtues traditionally described, except that these will now show a much harder edge, reflecting the real power which women now hold.

These 'feminine values' are now being adopted by the *whole* of society. They emphasize co-operation rather than competition, with a softness of touch which hides a real grasp of power. Indeed, the real power of the new women will derive from the fact that they are naturally at the heart of the new society, while many men, especially those mindlessly committed to machismo, will be peripheral to it.

It is even arguable that, over the next century, the traditional position may be reversed, with women holding marginally more power than men, leading to tensions within some parts of society. As we have seen, one of the basic building blocks of society, the nuclear family, is fragmenting. Where this happens, the one strong tie which remains is that between the mother and her children.

Now that there is the genuine alternative of divorce, where the law now recognizes this as a right in most Western countries (a relatively recent development), and there is increased prosperity (backed in many countries by some form of welfare state), there is now a real choice for women. They can consider courses of action which were previously not affordable, and could not be considered even in the most disastrous marriages.

Women as the New Managers

The growing power of women will not be limited to their traditional role in the home. Although their recent advances into the world of work have been in lower-paid positions, it is already evident that they will increasingly take on their male counterparts for the best jobs – especially those in the lower levels of management. While they may yet be held back by the traditional barriers to their progress, the infamous 'glass ceilings', they will ultimately crash through these in ever-larger numbers as the twenty-first century progresses.

Indeed, it is arguable that they are now better placed than men to answer the new challenges to management posed by the changing environment. The evidence is that, on average, women are more intelligent than men (at least in terms of measured IQ, whatever that might mean), albeit marginally so, and this is an important factor in the new knowledge industries. This effect is magnified by the fact that girls are more willing than boys to devote their energies to basic education, where the latter often underachieve in their teens. The net effect is that most women leave education, and enter the workforce, with better track-records; and are accordingly more attractive as employees – especially as potential highflyers. Further, in the new atmosphere of management, where co-operation rather than competition will be wanted, their traditionally less aggressive approach will also tend to make them better managers than their male counterparts. The reduced levels of aggression which may ensue will be better for all of us.

It is likely, therefore, that there will be a growing proportion of women managers. This will inevitably result in some tensions with the male managers who see their positions being weakened and their career prospects destroyed. It is not clear how such tensions may be resolved. The last management roles to succumb are likely to be those in senior management. More important is the belief, held by the combatants at least, that higher management positions demand more aggression than women tend to deploy. Above all, though, the establishment, which is predominantly male (and proud of it), is likely to close ranks, creating tensions among women managers.

One major driver for social change will be the growing power of women. Despite the low starting point, these changes are likely to have significant impacts -- with women taking dominant positions in some sectors. As a result, the twenty-first century may eventually come to be known as the 'women's century'.

A key element of the growing female power is ownership of the family. They now are gaining the sole right to decide the future of their children – often over the father's wishes – which gives them significant power over the future of society. In addition, legal and financial factors no longer disadvantage the woman seeking divorce.

The evidence is that women are, on average, better suited to the new management roles than their male counterparts. This may, however, produce tensions and the highest-level glass ceilings are likely to remain in place for some time, as women are still barred from senior management.

As a footnote, it should be noted that the relative brevity of this section about women comes about because, for once, the issues are clear cut; and, in terms of future development, quite predictable. It should not be seen as reflecting any lesser importance for these topics. The importance of the growth in power, and especially of influence, held by women in the twenty-first century cannot be overestimated.

Ethnic and Religious Tensions

So far I have not addressed the problems traditionally encountered by the other minorities even though these have seemed, over recent decades, to offer one of the greatest threats to global stability. These many minorities have long been persecuted for their religious differences, for their racial origins or just the colour of their skin. Discrimination in general, and racism in particular, have disfigured societies around the world, led to the deaths of millions and suffering for many millions more; hence the historical expectation that such racial tensions might continue for ever. It would be foolish to say that these problems will disappear overnight. Bosnia and Rwanda indicate just how serious these may become locally, and Northern Ireland shows how they can last for centuries.

Yet, surprisingly, the evidence is that the stresses which lead to the worst excesses seem to be *reducing*; apart from the growth of fundamentalism, our groups placed no significant emphasis on specifically ethnic problems. It is worth noting that most of the current crop of problems which are supposedly 'ethnic' in nature are in fact most closely associated with a history of 'colonial' exploitation: the English in Ireland; the French in Rwanda and Algeria; both the USSR and the US in Somalia. It is not surprising, therefore, that those disadvantaged in this way would wish to overturn their oppressors. Possibly the main reason for the resulting stresses we see breaking out in violence now is that the Cold War, which indirectly funded many of these episodes of exploitation, is now over. The covert forces of the USSR and the US no longer need to buy allies in this way and, without their support, the local regimes are at last being forced to, literally, come to terms with those whom they have previously suppressed with impunity.

The one remaining danger is that the Western allies, especially the US, feel that they must intervene and once more bring their own conflicts into the process. The more they get involved – in Bosnia, for instance – the *worse* the problems become as their own remote equivocation gets the better of local diplomacy.

The fact that these events are occurring now, no matter how regrettable they are individually, should paradoxically be seen as a hopeful sign. It indicates that the forces of repression are actually being rolled back across the globe; and if the Western power bloc can avoid the temptation to force its own solutions (and problems) on the participants, we should see fewer and fewer outbreaks in future. More importantly, within mixed communities the tensions should also reduce. One of the less obvious results of the growth in individualism, and the resulting breakdown of group identities and loyalties, is that differences between such groups will become less evident. They will be seen as just another aspect of the many lifestyles which will flourish. In particular, ethnic lifestyles will no longer be seen as a challenge to the host culture; in many respects there will no longer be any such thing as a single host culture. This amelioration of ethnic tensions will allow such groups to develop their own lifestyles unhindered, even if this does result in a voluntary form of ghettoization. On the other hand, it will eventually also dilute such ethnic groups as their members, no longer motivated by shared adversity, are in turn imbued with the desire to develop as individuals. The result will be a new form of cultural absorption, one which is potentially much more positive in nature since its converts will be free for once to take the best of their ethnic culture into the new melting pot with them.

The breakdown of group identities, and with them the disappearance of many group loyalties, may prove to be an even more important force for change than we have allowed for. So far we, and especially our politicians, have only talked about the problems this is causing. Perhaps we should instead look at some of the major opportunities it offers.

Third World Riches

On the global scale, by far the biggest group of underprivileged now are located in the developing world. As an aside, it is worth noting the change in terminology itself. These nations used to be collectively known as the Third World, a term I still prefer to use since it nicely conveys the historical attitudes which still can be important in understanding Western approaches. The implication was that these benighted nations were consigned, by history, to be forever excluded from any of the riches the West enjoyed. Now, as *developing* countries, it is assumed that they are already on the ladder of development – no matter how low the rung they occupy – and will ultimately enter the club of developed nations. The accelerating pace of development is quite startling. Britain needed more than half the nineteenth century to double its real income per head, as did America. Japan took the first third of the twentieth century. But South Korea and, more recently still, China have done it in a decade. We may thus see most of the Third World becoming relatively prosperous in much less than our lifetime.

The really big change in the distribution of global wealth and power will be the emergence of these 'Third World' states ultimately to become the dominant economic and political force. It has long been claimed that there should be a convergence between nations and the gap between them will narrow. In practice this has not happened until recently; now this convergence is being observed.

Unfortunately, as yet, few in the 'West' have even considered that their own days as 'top dogs' are numbered, and even fewer – if any – have worked out strategies to deal with this sea change. The reality, when it arrives over the next few decades, may come as a considerable shock to many of its leaders. Though it is unlikely to affect our comfortable standard of living, it may come as a shock to us personally. It will undermine the pride we still have in our nation; flag waving still has a very emotional appeal!

The Pacific Rim countries have already made their dash for growth to emerge as major challengers, in economic terms at least, to the West. For instance, China's Guangdong province, helped by Hong Kong management, finance, technology and marketing, has achieved an average annual real growth rate over the past quarter-century of more than 12 per cent. The South American countries also seem to be some way along the same road now that the US has reduced its destabilization campaigns in the region and, with the creation of NAFTA, are making encouraging noises. This leaves Africa, which is only now starting its own climb out of the abyss and which still seems, at times, perilously close to falling back into it. Of its two main potential economic super-powers, South Africa is, at long last, on its way to full economic development. The other, Nigeria – potentially by far the richest on the continent – is, if not quite in a state of civil war, still tearing itself apart economically and politically.

When both these governments get their act together, Africa too will become a powerhouse of change.

This particular aspect of income redistribution will have by far the greatest impact on the world as a whole. Western politicians continue to ignore developments in the Third World, but increasingly they do so at their peril. If there are to be ten billion worthwhile consumers by the early part of the twenty-first century, rather than the one billion the West now addresses, those in charge of the various Western economies – and those in charge of their multinationals – had better allow for this fact before the twentieth century ends.

A Genuinely Better World

In summary, then, twenty-first-century society will offer a much better environment for individuals to flourish. As we saw in previous chapters, it will certainly provide a richer base from which all individuals can develop their full potential. It will also be a much less divided world. But this will come about not because of any ethical considerations but because the have-nots, especially women, will grasp the power their numbers justify. In the process the rich will, reluctantly, come to accept – and proclaim – the virtues of such fair treatment; not least because they will fear that, as the new minority, they will in turn become the subjects of revenge – as have members of the establishment in previous revolutions. In this they will probably be wrong. A new spirit of co-operation, the most obvious sign of the feminization of society, will be extended to them too. It will, thus, be a world which is based on much more acceptable values, as well as a richer one. At long last, we will be able to have our cake and *share* it with others – and will probably enjoy the added glow of self-respect that this will bring!

Even so, the philosophy of individual empowerment is likely to be the greatest force for change in the next Millennium. Indeed, the essence of post-modernism, which is supposed to encapsulate these changes, now seems to revolve around the developing freedom of the individual to create his or her own lifestyle(s) – without regard to the groups which would traditionally have exerted (cultural) control over them. Related to this issue, in terms of the new resources individuals will need to have available to them, basic education will soon extend – worldwide – to ongoing, lifelong education. This will be especially important where, in changing times, those workers who are not trained to meet the changed needs will be unemployed.

The growing need for personnel to be multi-skilled, and often to adopt creative approaches to their work, means that self-management will increase rapidly; and organizational structures will need to change to reflect this, ultimately in some cases – especially in the rapidly growing fields of consultancy – moving to peer-to-peer collegiate relationships, or – to cope with the flexibility needed in the workplace – cellular organic structures which continuously adapt their internal and external relationships, led only by the culture.

One especially important driver for social change will be the growing power of women. Despite the low starting point, these changes are likely to have significant impacts – with women even taking dominant positions in some sectors. As a result, the twenty-first century may eventually come to be known as the 'women's century'.

The much lamented demise of the nuclear family – and of its values – is possibly less important than at first sight it might appear. New forms of the extended family seem to be developing, but a key element of the growing female power will still be ownership of the family. They now are gaining the, sole, right to decide the future of their children – often over the father's wishes – which gives them significant power over the future of society.

Outside of the family, the evidence is that women are, on average, better suited to the new management roles than their male counterparts – and will increasingly come to occupy the majority of junior and middle management positions. On the other hand, the highest level glass ceilings are likely to remain in place for some time, as women are still barred from senior management. The resulting tensions will prove uncomfortable – for both sides – in the short term.

6

AND JUST A FEW LOSERS

One of the factors breaking down group loyalties is the growing weakness of the individual's commitment to his or her work, where previously this often represented the prime (group) focus. The result is that whatever links might previously have tied them to their work-group, these are now regularly broken. Beyond these work-related changes, however, education has made individuals aware that there is life outside the narrow confines of the workplace. Their work is, in any case, still too often undemanding, bringing little fulfilment to their lives. They have, instead, turned to new interests into which they pour the energy and commitment which they formerly reserved for their work-group. It is in the context of these new interests that group loyalty, such as it now is, may increasingly be found: from fishing clubs to single-issue pressure groups. Whatever the reasons, employers can no longer expect their workforces automatically to make work the central element of their lives and that poses major challenges in terms of worker motivation.

Group identity is also suffering because individuals are becoming increasingly self-aware. Where previously they were willing to defer to their betters and, in particular, to group decisions, they are now aware of their rights and of the power they hold over their own future. One result is that they are much more willing to shift their allegiances to another group which they think more closely matches their own interests on a specific issue. This position has recently been unwittingly reinforced by governments, which have encouraged the breakdown of the strongest groups – such as unions – which they saw as opponents. Governments little realized that, in the process of pursuing their cherished ideologies, they were probably undermining the group loyalty of even their own supporters.

Despite the many reports about the growing greed of the masses, it is not necessarily true that these shifting allegiances reflect selfish interests. In an age of affluence, they may often have altruistic motives: a commitment to animal rights or the green environment may be just as much in evidence. The sight on UK television of middle-aged, middle-class women lying down in front of lorries – be they involved in building unwanted new roads or carrying badly treated animals – offers an interesting perspective on some possible future developments which might just give governments pause for thought. What happens when these same people move to causes which are genuinely revolutionary?

Work is no longer the prime focus of most individuals' lives – and that is resulting in a breakdown of group loyalties, as well as posing problems of worker motivation.

The growing self-awareness, and self-confidence, of individuals has resulted in them shifting their allegiances between groups to match their self-interest. This has weakened groups, and the social processes which depend upon them.

What, and Where, is the Community?

In this context, local communities will develop important new roles. These roles will be rather different from those of earlier communities. Not least of these differences will be those relating to the exact nature of the community. In earlier times this was easily defined, usually in terms of its geographic location, constrained by physical boundaries. Now, individuals may still belong to an identifiable local community – a village, say, with which they have strong links. This is, however, less likely; even close neighbours become relative strangers in our rapidly changing society. The *intangible* communities are even more disparate.

Many 'communitarians', led by Amitai Etzioni from George Washington University, seem to wish that society returned to its roots in the earlier, geographically based, communities and, especially, to the values these represented. Accordingly, some people are committed to creating a new moral, social and public order based on restored communities. *The Economist* (24 December 1994) suggests, 'There is some truth here, albeit distorted and exaggerated. Many people share the vague sense of loss to which communitarians appeal ... [but] communitarians prefer not to look closely at the past we have lost; their appeal is partly to nostalgia, which can tolerate only so much analysis.' Thus, while communitarians' objectives are laudable and they may offer useful ideas in terms of the ideal values of future communities, unfortunately they are somewhat short, as yet, of practical explanation as to how these could be made to work across the whole range of different communities.

The emerging communities are, instead, those of shared interest, with fellow enthusiasts; of shared spirit, with others aiming for similar fulfilment; of shared action, with those seeking quasi-political outcomes; of shared knowledge, with fellow scholars; and so on. It is too early to predict whether those meetings of minds – typically now in cyberspace, on the Internet, say – will develop any of the power of more traditional group contacts. Thus, new physical communities may eventually be created by those who have already found shared interests on the Web; the reverse of traditional links. On the other hand, research suggests that, as yet, the most successful computer conferencing takes place between those who also meet face-to-face – exactly the opposite of what is expected!

One problem which may occur, caused by the physical remoteness of other members of the group and the ethereal nature of the contact with them, is that these links will be relatively weak and, as such, potentially much less satisfying than the traditional ones. No matter how close your shared interests, an electronic shoulder to cry on does not offer the same comfort as a real one! Neither will such clinically cold relationships normally be able to offer friendship, or even a real sense of fulfilment. At the same time, the number of these remote groups claiming the loyalty and the time of the specific individual will grow, posing problems of management. The complexities of addressing the many options on offer may disconcert not a few of them.

There will, therefore, be an emerging role for someone, or something, to act as an intermediary beyond that offered by impersonal agents on the Web. This may be only in terms of advising on and, to a degree, helping to manage the individual's portfolio of options. Providers of ongoing education may be one source of support here. From a different direction, the social services, or even leisure departments of city councils might equally offer themselves, as might commercial organizations, once it becomes obvious that this will be a large (and potentially very profitable) market. Replacing the support offered by the physical community of old is becoming one of the major demands on local-government resources over the longer term.

The most important gap which is emerging is that of *personal relationships* within these groups. The problem is, as we have seen, that the group structures which used to support the richest relationships are rapidly disappearing. Something needs to replace these vanishing structures, before the community as a whole breaks down. The solution may, in part, come from the individuals themselves, who will need to learn how to develop friendships much more rapidly, at a shallower level, and then relinquish them – or maintain them only through cyberspace – as the people involved move on.

In terms of society's answer to the problem, unfortunately it is not obvious from which directions any developments might come, though it is likely that they will revolve around a range of solutions rather than focus on just one archetypal new form of community. It will be interesting to see what forms of organization do emerge to fill the gap: perhaps a virtual reality soap will satisfy some, allowing them to chat to a (seemingly) caring neighbour; maybe we will all take up pets and religion, just for the company.

In general, this is as yet one of the great unknowns. Our personal relationships are usually the key to our personal happiness. We know that the frameworks within which these relationships operate are breaking down; but we do not yet have any firm indications as to what will replace them. In many areas more pluralistic relationships do seem to be replacing at least some of them and, as we will see later, it is often the societal value system which undermines these new developments.

It is likely that the individual's traditional links to geographical communities will continue to diminish in importance, despite nostalgic remembrances of the values they held in earlier times. The groups to which individuals *do* belong will grow in number and in physical separation as contact increasingly moves to cyberspace.

The new, cyberspace, group relationships are likely to be relatively weak, offering little real fulfilment or friendship. They will also be relatively large in number, possibly disorienting some of the individuals trying to use them. It is likely, therefore, that intermediaries will be needed to handle, at least, the complexity of the individual's portfolio. People will still need friends! How this gap will be filled, by what form(s) of organization, is not yet clear; though networks of more pluralistic relationships do seem to be growing.

New Values

Despite the nostalgia for lost values, it is likely that the new organizations, even those providing group support at the personal level, will be essentially value-free. Some basic values will of course remain, but these will be the core values generally espoused by society as a whole: fairness, sympathy, empathy, charity, etc. These are precisely the values espoused by the communitarians. They will not be the old, loaded values previously demanded from group members: patriotism, group solidarity, etc.

Values will increasingly become a private matter. Almost all our groups felt that this would be the case. What the new values might be has yet to be established.

Our own groups clearly had difficulty with this. Their interest in religion (half of the groups) might also have reflected this issue, since their views did seem to identify a real need for religious values to be resurrected. Commercial organizations might seize the opportunity to create and own this local interface, but this is unlikely to be undertaken by firms steeped in the traditions of the 1980s. As will be discussed in more detail later, the new model for the firm will probably be based on co-operation not competition, especially with regard to customers or clients.

It is arguable that the commercial organizations which may succeed will appear more like religious organizations than other commercial firms. It was interesting to note that, as early as the 1960s, Hugh Hefner's approach to setting up the Playboy organizations – one of the classic exercises in lifestyle marketing – was driven at least as much by a quasi-religious belief in the values it incorporated as by commercial acumen.

In the developing commercial world, this move from competitive combat-driven to co-operative value-driven approaches is likely to accelerate. As we have seen, it is much more in keeping with the demands of the emerging markets. It is also much more in keeping with the demands of employees who, given the choice of jobs and the more than adequate salaries which will soon be on offer (as the recessions of the late twentieth century finally come to an end), will choose to work for an organization whose values they respect. It is now becoming the philosophy once more being adopted by leading-edge governments, most notably by what is becoming the pre-eminent supranational body, the European Commission.

The key symbol of the change is likely to be a move away from the commitment to money as the only worthwhile objective for both organizations and individuals. Apart from the US, where the philosophy is probably too engrained, the pursuit of money for its own sake will probably decline. Certainly, flaunting it, not least in the forms of conspicuous consumption which were so much in evidence during the 1980s, will probably come to be judged to be in the worst possible taste. This will have implications for the wealthy. They will increasingly act against those who threaten to expose them – those who allow themselves to be seen as receiving obscenely large sums of money, from whatever source – by ruthlessly sacrificing these to the mob. While the financial services sector will remain important, its values too will change, prompted by increasingly draconian regulation! Finally, the reduction of inequality, which is likely to develop as a potent political force, will also become a potent commercial force, especially for those large organizations which may be most at risk from political changes around the world.

Organizational Ethics

Even though we believe that most organizations will become essentially value-free, one of the most pervasive moral issues is at present that of the overall ethics of the organization, as opposed to individual morality which we have so far examined. The problem is that organizational ethics are, despite the proud boasts of the business community, in practice no better than sketchy. Badaracco and Webb, in their widely reported article (1995), write about how ethics are really implemented:

> First, in many cases, young managers received explicit instructions from their middle-manager bosses or felt strong organizational pressure to do things that they believed were sleazy, unethical, or sometimes illegal. Second, corporate ethics programs, codes of conduct, mission statements, hot lines, and the like provided little help. Third, many of the younger managers believed that their company's executives were out-of-touch on ethical issues. . . . Fourth, younger managers resolved the dilemmas they faced largely on the basis of personal reflection and individual values.

It is no surprise, therefore, that the Henley Centre (reported by Sheena Carmichael, 1995) found that 'At present in the UK, only 15% of the public broadly trusts multinational businesses to be honest and fair (compared with 27% who trust their newspapers, 39% who trust accountants and 83% who trust their GPs).' This is bad news for corporate ethics, but almost as bad for newspaper ethics! Three-quarters of individuals expect organizations to 'value ethics' within two decades. It may be that the revolutionary impacts of the popular rejection of political parties, and of the sleaze they are seen to represent, will be the main force that accelerates the rate of change in organizational ethics; and the mid-1990s' backlash is now targeted on senior managers in industry as much as on their political masters.

In terms of the overall values it espouses, future society is likely to become relatively value-free, though individuals will be free to take strong positions on issues which they personally judge to be important, and will do so.

The pursuit of money, driven by market forces, is likely to reduce in importance even for commercial organizations, many of which will become increasingly value-driven. Money itself is likely to become unfashionable in the face of regulatory control of its abuse and worldwide moves towards greater equality.

In terms of actual practice, organizational ethics will improve to reflect the high standards expected by employees, rather than the low ones set by politicians.

Changing Values

What is surprising is the fact that there has been so little recognition of, let alone acceptance of, the fact that a radically new set of values is emerging. Why has this been so ignored? The fact that there is a massive change taking place is obvious. Yet, in recent times, debate, such as it is, has almost always revolved around a loss of values. Around us we choose to see declining morals, breakdown of the family, lack of responsibility, no group loyalties, increasing crime, and so on. But almost none of us talk about the subject more positively, in terms of the new value sets which are emerging to replace those which are going into terminal decline. As we will see in a later chapter, such confusion is typical in the period of 'revolutionary' changes we are experiencing.

People have been through periods of significant changes in values before. These previous examples, however, were more obvious. They were typically the result of alien values being imposed by conquerors from outside the society: as the Normans imposed their own culture and language on the Anglo-Saxons in England; or by new groups gaining dominance from within the society, as the communists rewrote life in Russia. The difference now is that the changes are taking place within individuals across society.

The immense changes under way across almost all aspects of society, which are described in the rest of this book, are inevitably leading to equally immense changes in our value systems. In turn, the value systems are mediating the way in which, and the rate at which, these other changes take place. Perhaps the links between the two are best seen by looking at some examples covered in more general terms elsewhere:

INDIVIDUALISM Where so many of our core values such as the fairness, sympathy, empathy and charity mentioned earlier have emerged from our relations with the group to which we belonged, it is inevitable that new forms of these will be needed to cope with our newly empowered individuals, operating in loose peer-to-peer networks rather than ordered hierarchies. It is unhelpful, therefore, when politicians depict individualism largely in terms of personal greed. Fortunately, individuals seem quite capable of developing their own new value sets, based upon more sensible, and indeed more ethical, considerations.

COMMUNITY This also means we will have to renegotiate our links with the communities to which we belong, and to reshape and re-create (or create anew) those communities. Here, the attempts by our leaders to stem the changes to our existing communities again do not help. Much better to let the new forms emerge, even if we are, as yet, uncertain as to what they might be.

FAMILY The family is evolving into new forms rather than breaking down. This is where the establishment's defence of the status quo (the religious establishment as much as the political one) is especially damaging. The demonization of those pioneering the emerging new forms – be they single parents, divorcees or social workers – distracts our attention from the task of recognizing, and supporting, the new structures.

DRUGS As a much more trivial example but one which typifies the problems, the new drugs accepted by a majority of young people and seeming to have few impacts on society, are subjected to a greater degree of criminalization than some serious crimes against the person, whereas the existing drugs (especially alcohol and tobacco) are sold on every street corner despite the fact that they lead to many more deaths, of innocent victims as well as the users.

In the context of this book, therefore, perhaps the greatest impact of changing values comes not from what they are to become but from the friction between them and the old ones desperately promoted by an embattled establishment in an uncertain age. Many of our current problems and fears come from the attempts to hold back the tide of social development, demonizing, and even criminalizing, the new developments. If we see the loss of old values simply as a loss of morality, then we will be nostalgically locked into a past we can never regain.

Social Evolution

We have long been made aware of Darwin's theory of evolution, and the human species is probably still slowly following some lines of physical evolution though, diluted by our desire to protect rather than destroy the weakest in society, the forces which previously led to the survival of only the fittest have now been significantly reduced. As we have already seen, evolution of the individual may now become a matter of personal choice (coupled with the ability to afford this) – in terms of symbiosis with the computer networks – though this in no way invalidates the importance of such new evolutionary forces.

The real evolutionary pressures have moved from the individual to the group, from the physical to the social; and the rate of change has escalated by many orders of magnitude. In just a few millennia we have seen our position change from something approaching that of the solitary aphid to that of a colony of ants, a process which took many millions of years in nature. The structures of our society have changed, and are still changing, at a remarkable speed to develop a richness and complexity which is now the trademark of human life.

Social evolution has long since taken over from individual, physical evolution as the main mechanism for the future development of the human species.

Heritage and the Distance from Nature

In a general sense, society is rapidly distancing itself from nature, even that tamed part of nature which is represented by agriculture. It is creating an artificial world, a construct which gives false values to history in the form of the heritage industry. This need not be a problem, since each culture views its history in terms of its own values and this helps the individuals within that culture to better understand their own position within it, if not the true facts of history.

It may seem self-evident, but it is still important to report that a growing proportion of humanity – not least in the Third World – will continue to migrate to the cities. In the developed world there may be some return to town life away from the cities with their growing problems, and even some developing countries (helped by the new information technologies) may be able to persuade their populations to avoid the cities. Even so, it is the vast sprawling conurbations which will continue to shape the environment of many individuals. This possibility of global urbanization is often viewed with alarm. On the other hand, cities are remarkably efficient machines, in terms of usage of the infrastructure, for supporting their inhabitants. The problem is the failure to provide the correct, properly funded, infrastructures; and it is on this aspect we might most profitably focus.

As just one poignant example of the more general changes which have come about, one has been the disappearance of smells – as opposed to fragrances – in general, and those of nature in particular. The emergence of a global society dedicated to cleanliness is one of the characteristics of our urbanized society. The equally savage attack on smells, in the house and on our bodies, would also have mystified our ancestors. The process is possibly most obvious when you follow the visit of a group of Western tourists to the Third World. The horror with which they greet dust blowing around earthen floors, and the farmyard smells which are not discretely hidden as in the West, is real. These are interpreted by them not as nature but as dirt and squalor, which repulse most Western tourists who see them as very real signs of backwardness. This rejection of the genuine facts of nature and the creation of an artificial, literally sanitized version is likely to continue. It will not have dramatic impacts on society, but the hidden values behind the heritage industry, as well as those of the detergent manufacturers, need to be recognized if we are not to misunderstand what it represents.

Control over the Local Environment

The construction of an artificial environment is perhaps most complete in the home, which is now not just our castle but our hermetically sealed capsule protecting us from the dirt and smells of the world outside. This 'castle' has become a very conservative institution; the forms of our housing have changed remarkably little over the past couple of centuries. This may be no bad thing if the end result is efficient and incorporates the lessons learned over these decades. Indeed, in many regions even the humble and much maligned mud hut works well since its thick mud walls keep the inside cool in the midday heat and warm in the night-time cold and can be repaired by simply slapping on another handful of mud! Yet the evidence does seem to point towards the need to introduce new forms of dwelling, using materials and technologies which are more appropriate.

This desire for control of the 'personal' environment has long been extended to the surrounding community: positively in terms of the provision of parks, neutrally in terms of the council cleaning services and negatively in the form of planning controls.

Dick Atkinson (1994) suggests that such things are even more important. 'After the home, it is the street and the local park with which children identify. If beliefs give a personal identity, place gives a bounded geographical and physical sense of belonging.' Fortunately, the creation of local environments, to match the needs of the local community, will probably continue and accelerate among those communities which can afford them; once more the underclasses, who most need their environment improving, will be excluded.

This regeneration, where it happens, may also regenerate some of the community links which are disappearing. It may even provide the underclasses with some temporary relief from unemployment. Most of the work created may be in the service sectors, demanding the unskilled labour which has few other outlets. In line with what was said earlier, it should be noted that this local environment will now normally be located within a city. The key aspect, which it shares with the heritage industry (and with shopping malls, which are now becoming the active heart of city communities), is that this local environment will be an artificial creation designed to meet the aspirations of each community.

Control over the Global Environment

This wider topic has been seen by the lobbyists working on behalf of the green environmental groups and in terms of media coverage influenced by their activities to be one of the most important issues for the future, a view seemingly supported by almost the whole population. In the developed world at least, we are now all supposed to be 'green'. Our research, however, has shown that the reality on the ground may be rather more complex. Indeed, we were surprised to find that less than half our groups even mentioned the subject, and, in our earlier research on industry scenarios, there had been a significant split between the third of managers who seemed to be firmly committed supporters of green policies and the remainder who were apathetic about them.

On the other hand, when asked directly, our individual respondents put almost all of the familiar green fears at the top of their lists in terms of importance, as the media reports might have predicted. Most important of all (at 9.0) were 'nuclear war' and 'global war', but – fortunately – only a quarter of respondents thought either was a possibility. In practical terms, the most important were seen to be 'global water shortage' (rated 8.8 with a 60 per cent chance of occurring before 2030) and 'serious overpopulation' (8.7, with an even higher 80 per cent, probability – also by 2030).

The key green environmental issue was that of global warming, the only one specifically identified by a significant number of groups (by two-thirds of them), coupled with fears of the resultant flooding (and, paradoxically, of a new ice age which might also result). Although 'global warming' was rated slightly lower (at 8.3) than some of the other environmental issues, it was seen as likely to occur by 80 per cent of individuals – and was predicted to happen much earlier, a decade earlier than the other environmental problems, by 2020. The fears of not so long ago, of the effects arising from the 'ozone hole', seem to be waning rapidly.

To quote Allen Tough (1995), professor at the University of Toronto: 'What is most important of all is the continued existence and flourishing of human civilization. The continuation of humanity, society, and culture.... No other goal, value, or priority is more important.' Rather more brutally, he puts in context the claims of some environmentalists: 'I do not agree with the view that each form of non-human life is equal in value and importance to human life.... Although I hope we can live in harmony with dolphins and elephants forever, I admit I would choose humanity's flourishing over theirs if I were forced to make a choice.'

We have now concluded that this 'selfish' set of beliefs seems to lie at the heart of the debate. It is right to take issue with those green-environmentalists who want to preserve the world in aspic. Most of those who took part in our research would, in line with their less combative outlook, wish no ill on any creature; but there would be a strict limit on the price they would be prepared to pay for this ethical stand. The majority of the general public still buy cheaper, battery-hen-produced eggs even though they oppose the idea in principle. In any case, the corollary of Darwinian theory is that species are constantly dying out to be replaced by others – a lesson which the general public understand rather better than the defenders of the crested newt.

Missed by the 'green' campaigners, however, was the point that when the 'green' picture is painted in terms of survival of our own species, this does touch a raw nerve. Thus, the greatest part of the population seems to worry about the global environment specifically in terms of the future of our own (endangered?) species! It is the future of *humanity*, not the crested newt (or even whales), which really counts.

Most of humanity, now living in cities, is distancing itself from nature – not least by rejecting its dirt and smells – and is creating an artificial view of nature, which is increasingly being encapsulated in the booming heritage industry.

Communities, which can afford the cost, will increasingly work to control their local environment. While these different environments will be increasingly varied, they will reflect the needs and aspirations of each local community.

They may also lead to benefits in terms of regeneration of local relationships, and some relief in unemployment.

In general, amongst a complex set of responses, green environmental issues may be seen to be less important for the long-term future than coverage in the popular media might suggest; and where they are considered, they are largely seen in terms of the personal impacts. It is the future of humanity which really counts.

Disasters

The one threat which dominated the thinking of most of the groups was that posed by global disasters. It almost seemed as if this was just a symptom of a much deeper fear which surfaced in terms of the specific issues which might be topical. Thus, the major fear at the time our groups were meeting was of Earth colliding with another heavenly body. This is a real possibility, though of very low probability, and these results possibly reflected their interest in the spectacular collision at that time between the Shoemaker–Levy comet and Jupiter. Epidemics, of the AIDS type but not specifically AIDS itself (which was most often seen as being eventually the subject of a cure), were another of the unpredictable and uncontrollable threats which worried almost all the groups.

Global (nuclear) war was not generally seen as a significant threat. One suspects, however, it might well have come top of the list a decade earlier, and the current position probably does represent an important change in attitude. Even so, Joseph Coates (1994) predicts 'widespread contamination from a nuclear device which will have occurred accidentally or as an act of political/military violence'. With so many warheads still deployed, and with so many others being decommissioned in poorly guarded arsenals (and yet others being clandestinely developed), this is not an unreasonable prediction – even if it is an unwelcome one. The crucial difference from a decade or so earlier is that this no longer offers a doomsday threat. The outcome will not inevitably be the end of humankind in a hail of thousands of H-bombs. The most likely outcome is that, even though large numbers of people may die (albeit counted in thousands rather than millions), this will reinforce moves for global peace and towards global government. Groups also worried about the numbers of small wars, though these were no longer seen to pose a threat to the whole of humanity.

Other natural disasters, such as major earthquakes in key areas (including California and Japan, even before the Kobe disaster) were also seen to be important but not threatening global destruction. The fear of disasters also seems to fuel a degree of conservatism in regard to more revolutionary (technological) development, especially in the field of genetic engineering.

Most important of all, for the general groups at least, it seemed to link powerfully to the development of space as an insurance, a lifeboat, for the future of humanity.

Many of the (almost irrational) fears of the groups originated with threats from space itself: meteors or comets destroying life on Earth. Carl Sagan (1994), the eminent astronomer, reported that: 'Some 200 Earth-orbit-crossing asteroids and a much smaller number of Earth-orbit-crossing comets have been discovered.... There are thought to be some 2,000 objects as large as 1 kilometer in diameter, 320,000 as large as 100 meters and 150,000,000 as large as 10 meters.' He estimated that 'every 10,000 years one [impact of an object 200 meters in diameter] may have global climatic effects and every million years an impact tens of times more energetic than the aggregate yield of the world's current arsenal ... enough to cause a global catastrophe and kill a significant fraction of the human species.' Although such a collision is still a remote possibility, maybe those fears are not so irrational after all!

The solution to such threats is not just a lifeboat in space but to counter them from space. As Carl Sagan added, 'Proposals have been circulating since 1967 that recommend developing rocket and nuclear-weapon technologies to destroy or deflect near-Earth objects on impact trajectories with Earth.' Although nuclear weapons are currently banned in space, it seems likely this project will ultimately be undertaken – not least because it uses so much of the 'Star Wars' technology which would otherwise be wasted.

The main 'environmental' fear is of unpredictable and uncontrollable disasters, from space or from disease for instance. The main impact of this fear, apart from some conservatism about technological developments, will be a drive for the colonization of space as an insurance for the future of humankind, and as a platform from which to launch counter-measures against heavenly bodies on collision course with Earth.

Unemployment

Turning to a much bigger threat to society in the shorter term, unemployment still dominates much of the debate about the future of society. Leadbeater and Mulgan (1994) emphasize that 'the public regularly ranks tackling unemployment as its top priority. Yet the political world has proven unable to offer credible solutions.' They go further: 'a depressing conventional wisdom has taken hold – that high unemployment is inevitable and irreversible. It is based on impersonal global economic forces which are beyond our control or on the moral failings of individuals who are not prepared to look hard enough for work.'

The reality is probably more complex. Indeed, I would argue that, in the most important respects, the current large-scale unemployment is a temporary effect of revolutionary pains. The first Industrial Revolution displayed very similar symptoms, which we now recognize as being due to the 'structural' changes then taking place. Agricultural workers, no longer needed by a much more efficient farming sector, were displaced in their millions. Thus, the term 'structural unemployment', which is often bandied about, may actually offer a good description. The key, though, is that this is not permanent, it will be rectified once the structural changes are complete, as the displaced workers two centuries ago eventually found employment in the new factories.

Our view is, therefore, rather more optimistic. In any case, in the West at least, the demographic changes – with fewer young people entering the workforce – will ultimately result in a shortage of skilled labour. Indeed, the European Commission expects this to be the major factor in its labour market by 2005.

The major problem which remains, for the underclasses, is that the new employment will demand new skills. Some commentators talk about a new divide between those who have IT skills and those who do not, and it is the latter who will then make up the new, enlarged underclasses forever excluded from the wealth of society. No less than 80 per cent of individuals expect this to be the case within fifteen years. 'The nightmare scenario', according to Ian Pearson (1997) of British Telecom, 'is that society will be riven in two, with the employed information literate having as little as possible to do with the unemployed illiterates in the underclass.' Making a more general point, Leadbeater and Mulgan (1994) stress 'since 1950 five million jobs in the UK have gone from the goods-producing industries. Almost eight million new jobs have been created in services, both public and private. But this has not been enough to prevent unemployment rising, partly because the new jobs have not been taken up by men made unemployed [the denizens of the new underclasses] and partly because more women have taken up work.'

However, I suspect this is an alarmist view. The illiterate farm-workers driven off the land in the Industrial Revolution soon became the literate workers in the new factories. Our own computer illiterates may as soon develop the necessary new skills – especially when computer manufacturers are working so hard to improve the friendliness of their machines. The unskilled worker may by default gain as much from these developments in computing as the impatient manager, who the computer manufacturers are really courting.

It should be stressed that this is one subject where there are real regional differences. The reaction to much the same trends has been different in the UK and the US. The former has experienced the massive increases in unemployment described above. The latter has, in contrast with the whole of Europe, not just the UK, not seen similar rises in unemployment – but (with a more flexible labour market) has seen equally disturbing reductions in income for those in the underclass; and, as Leadbeater and Mulgan (1994) point out, 'the costs [in the US] have included growing wage inequalities, strong incentives for young men to go into crime and a huge increase in the prison population.'

Whichever the reason, these young men – locked into the underclasses – present a real problem. From the time they leave school they become just one of the anonymous unemployed (or underwaged); and they now are too often unemployable – not least because their raw strength, the source of their previous work and often of their personal pride as tough men, is no longer wanted by most employers.

Thus, beyond the lack of spending power, the spectacular disempowerment of these groups of underclass males, especially the younger members, is likely to result in major tensions between these groups and the rest of society. In effect they have been made outcasts from society and, as they see it, outlaws. They are, as one result, increasingly likely to ignore the laws of that society – and who can blame them! In one respect, these underclass men are assuming the role that underprivileged women have lived with for generations. The difference is that the women, conditioned by society, accepted their fate uncomplainingly. The men are likely to protest much more vociferously, sometimes with riotous violence.

The most important lesson, here, is that the major problems posed by the male underclass await not the victims themselves but the rest of society. Society, as a whole, will pay the greatest price, and it is the problem facing society as a whole which needs to be addressed.

The Underclass Threat

As we have seen in the earlier chapter, most groups of the traditional disadvantaged are set to join the rest of society, and share in the spoils of affluence, before they pose a real threat to the stability of society. This is a major change for the better. On the other hand, it is possible that this will not be the case for those specifically identified as belonging to the 'underclasses'. Again, it is worth repeating that there are major differences in this respect between nations. In the US the term generally describes ghetto inhabitants who are typically black or Hispanic, unemployed, poorly educated and dependent upon welfare – and usually with disrupted single-parent families. More importantly, in that country the underclass is seen by the majority to be different from the rest of society, as criminals or drop-outs who almost deserve everything they get. Something similar has more recently happened in Europe's major cities, but it comes from somewhat different roots – it is a result of long-term unemployment. It is less desperate in nature, and those afflicted in this way are not seen to be outside mainstream society. In both cases, however, there are no easy solutions to hand. Almost everyone agrees that the unskilled, and uneducated, who make up this part of the long-term unemployed, are likely to be doomed, by the social changes taking place around them, to a continuation of their misery.

As seen first in America, it comes to affect whole neighbourhoods, whole inner-cities, as those left with some prosperity flee; and the downward spiral, into which sink estates and communities, gets under way. Even if we never encounter them personally, we all now recognize the pictures on television: of vandalized properties, burnt-out and boarded-up houses, delinquent children hurling stones at the police. Such business as there is also retreats, and theft and violence increase as frustrated youths turn to drugs, and to crime which alone can pay for the new addictions. Paradoxically, the typical victims of this crime are not the rich plutocrats – who, it might be argued, have contributed to the causes of such poverty – but the criminals' fellow paupers. When there are inner-city riots, it is the homes of the poor and, in particular, the shops and businesses of the few who have stayed to serve the poor which are looted and burned. As a result, the rich, in general, and governments, in particular, can afford to ignore such riots, since they do not even inconvenience them or their followers. Even in terms of murder, the terrible statistics, in the US at least, show that in most cases it is blacks killing blacks, not blacks killing whites! The fact that most of us never have to face these problems personally makes it much easier for our politicians to pretend that they do not really exist. It was significant that, in the UK, the one riot which did cause the government to react immediately was (uniquely) a middle-class revolt, against a (poll) tax, that started to burn down office blocks in the centre of London!

This willingness by politicians to ignore what is conveniently out of sight may be a major element in the creation of a dangerously divided society. For, the problem of the new underclasses is one of the great challenges of the twenty-first century, for the West in general and for the US in particular. It just could be a temporary one, resulting from short-term revolutionary pains. But it is one for which no clear solution has yet emerged; and it holds an alarming potential for creating unstoppable civil unrest, even forms of civil war, between the two sides of society – debilitating both in the process.

Apart from crime, welfare benefits fund almost the whole of these societies. *The Economist* (30 July 1994) says that, 'Such is the perverse destiny of a European model of a welfare state, devised in expectation of universal full-time employment for men.' Even more depressingly, it goes on to add, 'Unfortunately, job-creation is something for which Europe seems to have lost the knack. The public sector . . . is stretched to the limit; and the private sector seems incapable of filling the gap.' This is one of the key challenges for twenty-first century politicians.

Middle-Class Fears

A further problem for most governments, with the significant exception of that in the US, is that the majority of the population see the existence of the underclasses as a failure by government. This offends most people's sense of fair play and, in an affluent society, they do not believe such inequality need, or should, exist.

Beneath the commendably moral position of the public at large, there also lurks an even more powerful fear. When disaster can be suddenly visited on anyone, they are understandably (if not totally rationally) afraid of dropping out of society into these very underclasses. Indeed this fear may not be totally irrational, where the spectre of unemployment in the 1990s hung over managers (and professionals) as much as their workers. Indeed, all of these symptoms may arise from the 'revolutionary' pains mentioned in the first chapter. As such, they may be 'temporary', lasting perhaps a couple of decades. Whatever the reason, the middle classes almost more than the underclasses themselves, and certainly more than governments, now want the best possible safety nets. They are already demanding this, and the clamour will grow in many countries – with the possible exception of the US where the process may be too far advanced (though that poses different problems, as we shall see later).

It will be in governments' interest to address these fears. Just one symptom, 'job stress', is thought to cost the UK up to 10 per cent of GDP annually through sickness, staff turnover, poor productivity and premature death. For the population to become more flexible, and more amenable to the changes which governments must introduce, they must feel secure.

Underclass Crime

Government, and society as a whole, has an even better, rational reason for finding a solution to the problem of the underclasses.

While crime is obviously not the sole prerogative of this group, the evidence shows that it is often associated with such deprivation, not least when the members of the underclasses seek to escape from their woes through the use of drugs. In addition, most crimes are committed by young men. Offences typically peak at eighteen, the age group which is now hardest hit by unemployment and harsh social policies.

Thus, the cost of dealing with the underclasses, and with the associated drug culture, is increasing rapidly. The cost partly arises in the areas of policing; some parts of the UK are now no-go areas for police, and whole cities are in the US. It also arises from the judicial and penal systems, which are both being swamped by the tidal wave of drug-related crime. Increasingly, it is appearing in the personal bills being paid by the individual: as rapidly growing insurance charges, but often as the cost of hiring private security firms to protect them and possibly through fleeing to fortified ghettos. The direct costs soon outweigh any savings which can be made by withdrawing financial support from the 'deprived'. The indirect costs, the fear induced in the wider population – and the resulting lack of flexibility in the face of change – are incalculable. The increasing costs (social as well as economic) arising from unemployment are progressively overwhelming those of solving it.

This is not a problem which market forces will, or can, solve. Indeed, the problem itself is actually creating valuable new markets, in insurance and private security for instance. Paradoxically, the work involved in those crime-driven industries is counted as an increase in Gross Domestic Product, not the reduction it really is! Nor will it be solved by taking a quasi-moralistic stand, that any law-breaker must be punished, if the cost of the punishment itself is undermining the economic base of society. It clearly cannot make good social sense to have to switch the investment from higher education just so that you can punish those whose drug-taking lifestyles you disapprove of, no matter how strongly you disapprove. There is little or no evidence that imprisoning large numbers of the underclasses, or of any part of the population, reduces the overall crime rate, and considerable evidence that it imposes significant cost burdens on society as a whole. A more enlightened, and certainly a more pragmatic approach to crime will pay dividends just in terms of the reduced costs.

Whatever the solution, one must be urgently found for the problems of the underclasses. In line with the earlier discussions about revolutionary pains, it is reasonable to expect that these (temporary) problems will eventually reduce as a matter of course. The underclasses will be re-employed, and indeed retrained, by industries which need them to meet a booming economy. In the meantime, the tensions may create structural distortions, even ruptures, in the fabric of society which will take much longer to heal. Looking on the bright side, this is a problem which most nations, outside the US, will not have to face; their electorates will never allow their governments to adopt such draconian approaches to the underclasses. Most will steadfastly oppose the idea of imprisoning as much as a third of the male population – as now happens in the black ghettoes of Los Angeles. On the other hand, the US as a whole, in which many of its states (not just California) are already well down the road to final rupture, may find it impossible to avoid the consequences. As a result, a long-drawn-out situation of simmering unrest, verging on civil war between the haves and have-nots, will disfigure many of its regions for decades to come. My apologies to my American readers, but this is one issue where the dark forces may win out!

The greatest losers from the growth in the power of women may be the male underclasses whose all-important macho image, already severely dented by the loss of their manual work, may be all but destroyed. The result is that they may increasingly move outside the laws set by the society which has rejected them.

Political demands for a solution to the problems of the underclasses will increase in most countries, partly for the sake of fairness but, most importantly, because of the fear they provoke among the majority of the population.

The cost of not resolving the underclass crisis is incalculable and cannot be long supported by most societies. Governments, with the possible exception of the US where the descent into the abyss may be too far advanced, are likely to be forced to invest considerable resources in the short term to remove the evil of the deprived, and the mounting costs of the related criminal activity in the long term.

7

PORTFOLIO LIVES

If we move on to examine individual lives in more detail, we find that this is where the greatest changes are likely to take place. We have already looked at some of these, in terms of their impact on society, and on the communities within this. Now we will examine how they will affect individuals such as ourselves.

The basic changes will come about because of the increased range of choices open to us. In part this will reflect increasing affluence: we will be able to afford a wider range of options. In part it will be because the provision of alternatives will expand as the markets respond to the new demand. So far this would seem to be in line with the increased hedonism we first observed in the 1980s; and there are those who would see this as the inheritance of the political approaches pioneered by Margaret Thatcher and Ronald Reagan. They would predict that Thatcher's 'children' might be expected to be greedy! In the main, however, it will come about because cultural constraints are removed or, more likely, discarded; the limitations imposed on individuals by the group culture will recede.

In reality almost all of us – across all sections of society, again with the exception of the deprived underclasses – will ultimately be able to build the pattern of life we would wish for ourselves. In any case, the richness of the patterns of these lives may be such that it will not seem unreasonable to describe them as a portfolio of different lives (or lifestyles) in a combination which is unique for each individual.

This new opportunity for individuals to select the portfolio of lifestyles which best suits their individual needs is one of the most important drivers for social change, not least in terms of diluting the links to their traditional roots.

The greatest freedom to emerge as the twenty-first century progresses will be that of individuals to choose the pattern of their lives to match their exact needs – unrestricted by group culture. The new richness of these lives may be best described as portfolio lives – a unique combination of lifestyles for each individual.

Post-Materialism

The richness of portfolio lives will come about at least partly because of a rapidly growing range of options of products and services. The choices will not be confined to those now available within each existing product category. Not least because the products of the whole world are now available in the local high street, recent developments have been mainly aimed at bettering existing product offerings. Thus in the UK, instead of tinned fruit in mid-winter, we can now buy fresh grapes from Chile or strawberries from California. This undoubtedly adds some richness but does not really amount to a revolution. Instead, it is arguable that it represents an incremental development on what exists, the final peak of materialism. The full richness emerges as people start to extend their overall buying patterns led by the expansion of the service sector.

The real consumer revolution – post-materialism, which represents a revolution in the buying behaviour of the individual – has been predicted by leading-edge marketers. These have claimed that individuals will slowly move from the acquisition of material things to the development of their lives to bring fulfilment. Even the OECD (Stevens and Michalski, 1996) admits that '... important changes are taking place in people's views about work. They point to a fundamental shift in attitudes within advanced societies away from materialist values and towards what have been described as post-materialist values . . .' and Michael Lerner (1996) reports that in his stress clinic, 'We were surprised to discover that these middle Americans often experience more stress from feeling that they are wasting lives doing meaningless work than from feeling that they are not making enough money.' In this context, 60 per cent of our individuals expect materialism to be replaced by ethics as the major driver (rated 6.3) in this field, by 2025. The marketers' approach, focusing on the post-materialist experience of individuals, contrasts with that of the sociologists, who have concentrated on post-modernism as a revolution in the way society structures itself.

Post-materialism represents a shift by the individual to a life, and related buying patterns, which addresses the inner self (by self-expression) rather than the outside world. The new values are supposed to be those of mature, well-balanced, caring individuals. I hope that this will be the outcome; but I await the evidence to back up this hope. When, and if, it does come, it will be driven by an individual's desire for genuine fulfilment rather than the self-gratification which the purchase of goods brings – there is a limit to how far simple materialism can go.

One problem is to decide exactly what is 'fulfilling', even for ourselves. It is almost impossible to determine what this might be for another individual! This was inevitably one of the problems experienced by a command economy, where the communist bureaucrats decided what was best for everyone. In the Western alternative, many large organizations, such as Disney and McDonalds, are happy to do much the same for consumers in general, with little regard for any responsibility for the outcomes, apart from profit. One genuine option is to educate all of us to make our own choices and then, as far as possible, to provide the resources to support these choices. Where the indications are that these choices will typically lead to inner development, rather than outward self-gratification, it is likely that the extra resources needed will be moderate in scale, adding weight to the earlier argument that we can now assume resources are effectively unlimited (since the demands on them will reduce rather than increase).

This leads, again, to the need for ongoing education, for improved skills in self-development. This recognition of our own individual needs, the core of individualism, has already emerged on a relatively large scale. The lifestyles debate in marketing has introduced the elements of 'inner-directedness' (self-fulfilment) in addition to that of 'outer-directedness' (hedonism). These have remained largely unrecognized among the population at large.

The right-wing politicians of the 1980s, most notably Margaret Thatcher and Ronald Reagan, talked a great deal about individualism, in general, and of freeing the individual to achieve his (or, less obviously, her) full potential, in particular. This was seized upon by the electorate, whose emerging aspirations it matched; and was acted upon by many of us, often swayed in the direction of hedonism by the political rhetoric and media comment prevailing at the time. The problem was that, while we were developing a genuine commitment to individualism in its wider sense, the politicians were misusing the term. In their eyes it was simply new rhetoric used to justify a continuation of their traditional positions. These politicians, where they said that individuals must be free to pursue their own ambitions, did not mean all of us. They meant, instead, just those fortunate owners of capital who already dominated society and the ownership of wealth within it – now being given even more.

At a time when almost everyone was becoming wealthier, at least outside the US, the switch of more of this to the already wealthy was not generally noticed; and the rhetoric of the politicians was for a while accepted at face value. When the recession hit, at the end of the 1980s, the duplicity of the politicians was revealed to all – though, once more, the picture was confused by the fact that the politicians in opposition, as much as those in power, failed to recognize what was happening.

The key lesson, however, is that the move to individualism (as opposed to hedonism, pure and simple) has been under way for some time, encouraged initially by opportunistic politicians who did not recognize the forces they were unleashing. This trend poses a challenge, for the individuals as much as for governments. In the past, most of us have been very limited in what we could expect to make of our lives and were, on the whole, content with our limited lot. Much of our time was spent in just surviving: in subsistence agriculture and then in the dark satanic mills. Only for the élite, and especially for the aristocracy, was there some choice. For most of us, though, it has only been during the past few decades – and then only in the West – that increasing affluence has allowed us even limited choices.

The twenty-first century, from its beginning in the West and later for the Third World, will see our range of choices extend dramatically. As we saw, in part this will result from increased affluence, allowing the sort of hedonism the establishment now expect us to pursue. In the main, though, it will come about precisely because we will *not* choose hedonism, but will move to post-materialist choices (and values) which have less resource-intensive implications. Accordingly, governments should welcome the emergence of such values with open arms. They will allow a significant growth in 'consumer satisfaction' for a relatively insignificant drain on resources – just the sort of non-inflationary solution politicians are searching for.

The first challenge for individuals is to obtain the education necessary to discover and evaluate the choices on offer. For society, however, the challenges are more complex. In the first instance, there is the task of providing the infrastructure, not least the new forms of education, to meet these emerging demands from individuals. The result will also represent a fragmentation of society, at least in terms of the various demands it is making.

The development of the mature inner self, leading to inner peace and the end of the cycle of reincarnation, has long been the goal of the Eastern religions derived from Hinduism. This should be contrasted with the much more combative origins of Christianity.

As Karen Armstrong, once a Catholic nun, forcefully points out (1997), 'The God of Moses was a god of war. He was Yahweh Sabaoth – "Yahweh of Armies". Murderously partial, he sided with his own people, the Israelites' – a legacy which still stalks the Middle East. The noble sentiments of Eastern religions, expressed with particular force in Buddhism, have in general represented unearthly desires. Most individuals, even devotees of these religions, have been too busy surviving to devote much time to achieving such inner peace. Now, however, time and resources *are* available; especially in the West, where many of us are increasingly intrigued by the less combative values of the East.

There is a barely discernible trend away from purchasing goods for self-gratification, and the search for economic security, towards a quest for self-fulfilment and self-expression encapsulated in the term 'post-materialism'. It is likely that this trend will escalate rapidly through the first decades of the twenty-first century.

The adoption of self-fulfilment as the prime individual objective will require a major commitment to ongoing education of a new form, demanding significant new (interpersonal) skills from the educators involved.

Increasingly, individuals will demand full control over their own lives. Growing affluence, and the move to post-materialism, will make this a real possibility. It will, though, pose challenges for them, for society and above all for the politicians who will be isolated by the process.

The logical end of self-fulfilment may be, for many, the development of 'inner power' – or inner peace, a goal which characterizes Eastern religions. Interest in these religions may, therefore, grow, at the expense of the more combative Judaeo-Christian religions; and inner power may become a potent force in society as a whole.

Lifestage Roles

Moving on to a topic which may be just as important for us as individuals, society as a whole has traditionally recognized only a few effects of age. It is only in the past century or so that even children have been recognized as a separate group, as have those at the other end of the age range – pensioners in retirement. Within the working population there has been little differentiation.

Among marketers, however, the importance of the age of individuals has become increasingly important. The emergence of the teenage market, only invented half a century ago, at one extreme and the grey market, an even more recent invention, at the other indicate just some of the changes which have taken place. The most meaningful categorization of consumers has proved to be that of lifestage – which parallels age but also takes into account what is happening to the individual. Thus, it reflects when individuals enter into marriage and when they have children, both of which typically change their lifestyles dramatically but are not necessarily tied to a given age.

Longevity and Retirement

Not least of the major demographic changes facing governments around the world, especially in the West and Japan, are those resulting from increased longevity. We are living longer, even without major advances in medicine; and the indications are that by the middle of the twenty-first century the average lifespan expected of healthy adults might reach one hundred years without major breakthroughs in medicine. Even more optimistically, perhaps, two-thirds of individuals expect the average lifespan to *exceed* 100 years as early as 2035! With the breakthroughs which are possible, in the light of recent research into the ageing process, our lifespan, again for healthy adults (excluding factors such as infant mortality, which drag averages down), could soon rise beyond the current ceiling of 120 years. Just as important, we will be healthier in our old age. This is no abstract concept, for many of those who are children now are likely to see not just the dawn of the twenty-first century but that of the twenty-second – and perhaps some of their parents will do so as well!

The response of most governments to even the start of this process has been something approaching panic – especially where, from the 1960s onwards, they had often promised that our retirement age would be gradually reduced. They, abetted by their actuaries, could only see the rapidly escalating cost of providing pensions for such a large proportion of us – an even bigger problem when it is realised that, unlike the providers of private pensions, most governments have traditionally funded payments from current (tax) income rather than from investments set aside for the purpose. More than 90 per cent of individuals are worried about a pension crisis occurring in the next twenty years or so. This, however, only becomes an insuperable problem if you assume the only factor to change is the lifespan and that our retirement age is immutable. This assumption is, of course, false. The Japanese, as so often leading the field, have already *increased* their retirement age. It seems inevitable that other governments will be forced to follow suit, with our retirement age perhaps increasing to seventy years by 2010. The European Commission – with a better appreciation of long-term trends than most national governments – certainly expects this to happen.

This increase in our retirement age will pose other problems for governments which were looking to a reduction in it as a way of reducing unemployment, and to those organizations, following a similar line, which have been persuading their staff to take retirement at fifty-five or earlier. Both will need to find alternative solutions! But this problem should be easy to handle, once we recognize the new age structures, and should enormously benefit society.

The requirements of individuals will increasingly differ, dependent on their lifestage. Governments will have to reconcile these different demands in the context of all the other competing demands within society.

Increasing longevity will mean that, by the mid-twenty-first century, adults – already born – will expect an average lifespan of one hundred years or more.

This will inevitably result in the retirement age ultimately increasing to seventy years or beyond.

Work Patterns

A related problem is that the traditional model of work is still tied to that of manual labour. Beyond sixty-five, this says, the muscles are incapable of meeting the harsh demands of hard labour. Of course, this is now a nonsensical requirement for most of us. Our hardest physical task during the day probably is driving to work, where we sit behind a desk for the rest of the time. Even hard labour is not impossible, where we are now fitter at all ages – though, again, it poses the greatest challenge for the underclasses. The real work requirement now is for intellectual power or for communication skills and these powers fade much later, allowing us to be capable of this sort of work well into our seventies and now even into our eighties.

The caveat is that we probably won't want to continue the same job we have already been stuck in, from nine to five each day, for the best part of half a century. We will almost certainly demand more suitable employment and, with a growing proportion of us falling into this age category, we will have the power to make our demands stick. As a footnote, it is worth noting that the existing work patterns are not necessarily God-given. Mulgan and Wilkinson (1995) record the surprising fact that 'The pre-agrarian hunter gatherers spent on 15 hours each week engaged in work – which was often more demanding of intelligence than most work today!'

The problem of the non-retirees highlights a wider issue, that of the very different work patterns which emerge at each different lifestage we go through. Some of these are widely recognized. Thus, retirement is already clearly seen as a watershed when, in the last two decades of life, the patterns change dramatically. The subsequent stages – active retirement, in good health, leading to sheltered retirement, where the individual needs some help, leading to cared-for retirement, where he or (more often) she needs substantial care (typically in some form of nursing home) – are much less widely recognized. Retirement is, thus, not the same for all who are now lumped into the category, especially where it will, in future, probably cover an increasing number of decades. This fact needs to be more widely recognized, for the demands of the sub-groups are very different.

At the other extreme, the first two decades are now generally given over to education, preparing the young to enter society in general and the world of work in particular. The nature of this education will change, not least to prepare them for the changes they will encounter later in life, but at least the stage is recognized.

Between these extremes, however, are our five decades or so of 'working life' within which the various sub-stages are too often lumped together without distinction. The twenty-year-old is expected to compete with the sixty-year-old for the same jobs, to the disadvantage of both and, since it exploits the strengths of neither, to the disadvantage of society as well. Of course, employers already go some way towards recognizing these different stages, or at least the increasing level of skills and knowledge, by rewarding the later stages (until recently) in the form of seniority payments.

Our first decade in employment is usually now seen as a time of exploration, in many respects a continuing, but different, form of education. We want to broaden our horizons, and are willing to take risks, not least by changing our jobs frequently. The next two decades are now typically those dedicated to raising a family (in a nuclear family or as a single parent). With more mouths to feed, greater resources are needed and, in particular, greater stability of employment is essential. Now very aware of our responsibilities, we work hard to establish a long-term career.

The last two decades of working life now often come as a pleasant surprise. While security (in terms of retirement pensions, for example) is still important, with two incomes – and no children to support, and perhaps even some inherited wealth – we have the time and the resources to develop what we see as more fulfilling roles in life. To achieve this new ambition we need education: first to help us understand what these roles might be, and then to prepare us for the one(s) we choose. Above all, we need more flexibility, perhaps even in the form of part-time employment, and may well need to change jobs and even employers, abandoning the long-term career which earlier had seemed so important to us.

It is this last stage of the working life, the two decades or more which may extend beyond the traditional retirement age for the non-retirees, which is currently least understood or recognized. Even the European Commission, though it recognizes the problem, as yet has no plans to deal with it. The 1980s saw employers even seeming to wash their hands of workers in this category. Early retirement was often seen as a more socially responsible form of redundancy. Yet this group contains some of society's most productive individuals, and some of the most important contributors to its development – not least because they have the experience, and the time, to stand back and examine the wider perspectives.

It should be obvious that ongoing education will play a central role in the development of all of us, to prepare us for our transition between lifestages. The ideal focus for, and source of, this education should be the employers since it will necessarily be employment-centred. Japanese corporations already see this as a legitimate role for them, as do the more enlightened employers in the West. It does pose a problem, however, for the many organizations in the West which do not see their role as going much beyond mere exploitation of their workers; and they certainly do not see it in terms of providing expensive education for them which has no obvious immediate benefit. Once more, therefore, it seems that government must step into this breech – as the European Commission, despite its lack of firm plans, expects to do.

An increasing number of workers will continue beyond the present retirement age, but they will demand, and get, more suitable jobs.

Society in general, and employers in particular, will need to take more account of lifestage differences by developing older workers to their full potential rather than casting them on the scrap heap. The development of individuals through the various lifestages must include a significant role for education. Ideally employer-provided, it is likely in the West to depend on governments for its provision.

Education

This section is specifically about the future content and delivery of education. That content is already changing dramatically, and will continue to be revolutionized in the future. Traditionally it focused in the primary school on the techniques of the 3Rs, and we still hear many appeals for it to concentrate on these. On the other hand, reading – in the new IT age – requires an understanding of icons as much as of words; and writing now depends upon the correct use of the spell-checker. Simple arithmetic has not really been needed since the invention of the pocket calculator. Beyond these techniques is the whole realm of knowledge, now available at your fingertips on CD-ROM or via the Internet. These are gross simplifications and distortions; but the scale of changes they indicate should already have had a dramatic effect on education in general. In fact, the conservatism of practitioners, and the scarcity of resources, has delayed many of the necessary changes. A recent survey we conducted showed that no major educational suppliers were providing genuinely interactive offerings, and most CD-ROMs merely had existing text-books dumped on to them, but it is likely that practice will catch up with reality in the near future.

In terms of the techniques, the need will be to match them to modern requirements: to teach data acquisition in general rather than just reading (though that will still be essential, where the symbolic representation of ideas by words is becoming ever more important in society); to teach analytical skills in general rather than just arithmetic, say (though numeracy, too, is an essential skill for modern society); to teach communications skills in general, not just writing (though that also has become ever more important, albeit via a keyboard rather than pen). The point to emphasize is that these new skills do not preclude the older ones. It is only in recent decades that illiteracy has become a debilitating handicap. The addition of the new requirements will make education more demanding, even more important.

Beyond basic skills, much of education has traditionally been concerned with imparting basic knowledge, the learning of vast quantities of facts such as the dates of historical battles or the formulae for chemical reactions. Since computers will increasingly be able to (instantaneously) provide each of us with the information we need, down to the minutest detail with miraculous infallibility, it will no longer be necessary to fill our heads with all the knowledge we might need in a lifetime. But we need to teach the frameworks of knowledge, so that the individual can find their way to the correct database, and then put the resulting information in context.

The delivery of education is changing just as dramatically, not least because the major cost of ongoing education involves removing the individual from employment so that much of it has to take place on a part-time basis. The teaching of knowledge, and of many skills, is often now most efficiently provided by some form of distance learning; 80 per cent of our respondents, for instance, expected to be taught at home by computer (in 2020) and more than two-thirds foresaw the demise of the traditional school (by 2025). According to the scenarios developed within the Open University, in future this may be, as many pundits predict, delivered through the individual's personal computer – typically by network providers. It may also be through traditional printed material, delivered by the mailman, since at present this may be provided at perhaps just a tenth of the cost of the more high-tech solutions and, for many teaching purposes, it is only marginally less effective. The key factor, whatever the exact delivery system, is that considerable expertise can be invested in producing the very best quality of teaching material. The Open University for instance, spends the best part of a million dollars producing each two-hundred-hour course, building into it the knowledge of world-class experts.

The efficiency of this form of teaching and the potential economies of scale are such that, once the break-even volume has been achieved, extra students can be added almost for free. It is ideal, therefore, for large numbers of students, such as will be experienced as part of the new drive for ongoing education in the West, and especially for expanding basic education in the Third World where the large numbers of potential students simply can't be matched by equally large numbers of trained teachers. Indeed, the Open University is already pioneering such programmes, in Ethiopia and Eritrea (two of the poorest countries in the Third World), where we are teaching literally thousands who otherwise would have no hope of a university education.

The exact nature of these distance-taught offerings will range upwards from the equivalent of self-help books, provided through the network but still often in text form (even if spoken to camera by the author, offering less than ten hours of tuition to a few hundred users and resourced on a shoe-string. Above this basic level, and perhaps forming the bulk of the teaching load, the mass of education which is currently the province of schools and colleges will be carried by programmes such as those now provided by the Open University (OU), each taking hundreds of hours of education to thousands of students. These programmes will allow the extension to ongoing education, which probably could not, on the large scale, be resourced in any other way.

The originators of these mainstream distance-taught programmes, rather than individual courses, are likely to be relatively large institutions – the OU, with an income of more than £200 million each year, may be fairly typical – though their local deliverers may well be the same schools and colleges as at present. In general, only these specialist course developers will be able to afford the million dollars or more needed for each course – and the many million needing to be spent across a range of such courses (the OU's MBA programme by itself cost more than £14 million to develop!) – though some smaller institutions with specialized knowledge may offer individual courses, and publishers may also deliver some specialized modules.

Even so, the key characteristic of the new forms of education will be their reliance on economies of scale to allow major investments in development of high-quality material.

The most important developments, completely new, are likely to come from the even greater economies of scale which will result from individual computer-delivered courses, rather than programmes of education. Each of these individually delivered courses will cost much more to develop though. They will be fully interactive software products, based as much on computer games technology as on any conventional teaching concepts. Indeed, the investment for each of these is likely to look much more like that for a feature film (costing tens of millions of dollars) and the unit price to purchasers will (again following this pattern) be relatively low. Accordingly, suppliers will need to sell to millions of customers, not thousands, to recoup their investment. The mass appeal needed to recruit such large numbers of users will almost inevitably push such offerings away from traditional institutions to the fringe of the entertainment business, which is why they are now often referred to as 'edutainment'. They will also compete in the mass entertainment markets, which means that the providers of these are likely to emerge from the mass entertainment media.

Managing the Student

I have not, so far, discussed face-to-face teaching, and it might seem as if this will no longer be needed. Nothing could be further from the truth. Almost half the Open University's costs arise from the provision of such face-to-face teaching, in the form of tutoring and student management. This is a higher ratio than for most other distance-taught courses, but it reflects the reality of the OU's long experience in the field.

With the routine chore to delivering information relegated to distance-taught packages, teachers will be able to concentrate on the most productive face-to-face aspects of individual tutoring, something no package can offer and something for which most teachers currently have little time available. The extra time made available to teachers will also enable them to manage their students, programmes of study more directly, though, even here, computers may have a role. In this way, students will be able to follow programmes of study uniquely matched to their needs, introducing the concept of portfolio education designed to allow each student – supported by ongoing lifelong education – to fulfil his or her ultimate potential.

The whole basis of education is changing. The new content reflects the new demands of an information society, broadening and deepening the skills and knowledge imparted to students.

The base workload of education in general, and especially that of the rapidly growing ongoing education sector, will most likely be provided by a massive expansion of distance education, within a decade or so delivered electronically, though initially – and more economically – in printed form, following the pattern established by the Open University. The new factor will be the provision of mass-market 'edutainment', developed – on much the same basis as feature films – by the mass media.

Face-to-face teaching will continue to expand, but in rather different forms. At one extreme it will concentrate on tutoring, to meet the specific needs of individual students. At the other it will evolve to take responsibility for managing the students' education portfolio, designed to allow each student to fulfil his or her ultimate potential.

Lifestage Marriages

A less obvious aspect of lifestage is the choice of partner with which to share it. The third decade of life (from the late teens onwards) is now typically spent promiscuously with a variety of partners, ostensibly looking for the right lifetime partner but also enjoying the experience. The fourth and fifth decades are, in theory, then spent with that partner rearing children. In practice, despite all the pressures of society to conform, this is even now only followed by a bare majority in some sections of the population where almost as many now choose to be single parents.

The older couples, who have survived the rigours of child-rearing are supposed to emerge from it still motivated by a shared, romantic love, sharing the same interests through the happy years at the end of their lives. The rationale for this, apart from the fact that the habit has taken hold and it is what is traditionally expected of marriage, is less clear. Indeed, the very practical relationship which was necessary for rearing a family need not necessarily lead on to the more creative ones which may then be needed for the development of fulfilling lives, even where interests are shared. Where the interests diverge, as is quite likely in the new emphasis on very individual styles of self-fulfilment, the existing relationships may not be strong enough to endure the resulting tensions.

There is an argument, therefore, that new marriages, or less permanent partnerships at least (where different stages of fulfilment may demand a progression of partners), may be just as needed for the fifth decade and onwards as they are for the earlier lifestages. This will become even more evident when the average lifespan exceeds a hundred years. Seventy years with the wrong partner is surely too much for anyone to bear!

Charles Handy, echoing one of the major themes of this book, describes 'portfolio marriages'. 'Everyone will live a portfolio life one day for part of their lives. Most people will match that with a portfolio marriage. A portfolio marriage is not a recipe for polygamy, a different partner each day or night, nor is it an invitation to serial monogamy, a sequence of husbands or wives. Rather it is a way of adjusting a marriage to the differing demands of a changing portfolio in life. . . . If the relationship does not flex in some way, it will break.'

This will place enormous strains on marriages during the transition periods, not least where the culture still fails to recognize the reality of the new developments and to provide the support which is needed. It is the inability of the existing culture, and especially of the establishment, to accept the changes in lifestyle which may be creating many of the problems we see. Once society adopts a positive approach to them, we should find that the new relationships are potentially richer than the rather sterile model of the nuclear family which they are replacing.

Networking

One of the most important changes is the emergence of computer communications networks. For a few aficionados it already represents the central focus of their lives; they spend hours every day communicating with fellow net users around the world. What matters is not what these few million computer freaks do, but whether their numbers will grow into the hundreds of millions over the next decades as their enthusiasm comes to be shared by a large proportion of the population as a whole.

Computer conferencing, using email, is proving to be a peculiar form of human communication. It is an artificial discourse, lying somewhere between normal face-to-face conversation and the mass media. It runs into problems at the two extremes of usage. If there are only a few messages, participation tends to drift away and it dies. At the other extreme, information overload can be a real possibility. With enthusiasts flooding such a conference with hundreds of messages every day, it becomes incredibly difficult to sort the wheat from the chaff. Using a suitable message-handling system on your PC helps, but you are still dependent upon the sender's ability to highlight, usually in a pithy heading, the important content and upon your own skills in discarding the irrelevant.

Once beyond the initial stages, the use of networking should spread rapidly. Once we have gained the basic skills, and the all-important confidence in our own abilities, we can cope with far greater complexity. Use will spread even more rapidly once the infrastructure allows a richer diet, and more user-friendly forms of presentation. In particular, once the widespread use of voice and video allow the process to come closer to the traditional methods of conversation, it will open up even to those of us who can talk, but not write, fluently.

One driver of the past half-century of the electronic revolution has gone relatively unnoticed, possibly because it fitted so well, and so easily, into society. Long before there was a personal computer on every knowledge worker's desk there was a telephone which was used, often to talk to people on the other side of the world in exactly the same way that is now exciting people's imagination about the networks. The real precursor of the modern network, therefore, was not the complex mainframe computer but the humble telephone system. When the computer networks are as easy to use and as widely accepted as the telephone, the new information age will have genuinely arrived.

Computer networking represents a revolutionary change in human communication. Once skilled, users will rapidly accept the networks into their lives as easily as they did the telephone, and the use will significantly enrich their lives.

The Infinite Network

Ultimately the impact of networks will go far beyond this, to become in effect an extension of our nervous system. Whereas the most important commodity now being traded is information, we will have at our fingertips (or even embedded in our brains) instant access to almost all the information, in whatever form, which has ever been recorded. Then, we will have the massive computing power to manipulate the data we select in whatever way takes our fancy, all for affordable prices.

Some of us will use this power to decamp physically, taking our work with us as some of my colleagues already do, spending their summers in the beautiful countryside of Tuscany or on the Greek islands. My real workplace is now in my home, where my computers are. Others will use a computer to visit the Himalayas on the other side of the world without ever leaving home. Some will use it to turn hobbies into businesses, creating feature films or very sophisticated documentaries from the myriad of sources available without ever touching a camera.

Geoff Mulgan suggests some totally new trades which may emerge: 'Internet plumbers (PC won't talk to your fridge? Call one out and they'll solve it) ... Workgroup synthesizers (bringing together ideas from staff from different projects in remote locations) ... System hosts (DJs and talk-show hosts of the Internet, who will be famous for the discussions they provoke each day – the true megastars of the future)'.

The power and freedom the networks increasingly offer will almost certainly accelerate the move for individual knowledge workers to become self-employed, at least in part. More generally, more than four-fifths of our individuals expect contract working, and multiple careers, to become widespread in the next twenty years. The ease with which almost any of us will be able to obtain and manipulate (information) resources, and then sell the resulting product to a worldwide market, should not be underestimated. In many knowledge markets this will put us, as individuals, on an equal footing with multinationals, especially where we have much greater control over our overheads or over what we decide to build into our pricing. There will, therefore, be every incentive for talented individuals to become self-employed, selling their wares (basically their own talents), on an ad-hoc or contract basis, in a seller's market.

This growing form of self-employment will be quite different in nature from the traditional forms such as those of the highly paid professional (dentists, for instance), at one extreme, and those providing low-paid personal services (for instance window cleaners) at the other. These new self-employed will offer what has been conventionally sourced inside larger organizations: white-collar administrative work. In the United Kingdom, for instance, when you ask for telephone directory services in London you are now quite likely to be transferred to an operator sitting in the living room of a croft on a remote Scottish island; but, even then, it will be a BT computer which undertakes the time-consuming task of reading the number to you!

Different individuals will approach the opportunities offered by the networks in different ways and, once more, the differences due to lifestage will often represent the most important impacts of their choice. Older employees, in the last two decades of employment, are the most likely to benefit. The existing trend for them to demand new roles, and often to take on part-time employment, will make these new opportunities very attractive for many of them and their employers. They will have the saleable skills built up over the years and the resources to invest in the new technologies and the necessary motivation. Some of those in the child-rearing stage may also benefit by enabling at least one parent to work from home, thus avoiding crippling child-care payments.

Individuals will soon have incalculable power at their fingertips, coupled with access to almost the whole of recorded human knowledge – in every form, and at affordable prices – and this will change their lives beyond recognition.

Networking is likely to stimulate a trend to self-employment, moving administrative work from inside organizations to outside of them by means of *ad hoc* contracts.

Major beneficiaries of networking are likely to be those older workers who are already searching for new roles and part-time employment. It also may offer some child-rearers the opportunity to look after their children while continuing in employment.

Teleworking

One aspect of networking, which has over recent years stimulated great interest but so far produced few practical results, is that of working from home in the romantically labelled 'electronic cottage'. In theory, networking is the ideal solution to this and the wider infrastructure will soon exist with fibre-optic cables due to pass close to most houses in the developed world. These developments are potentially capable of delivering easy access to the full power of networks to almost everyone in their homes. This is the superhighway concept at its most optimistic and, in view of the high levels of investment being made by the providers, it will probably happen very soon for many of us. For those BT telephone service operators sitting comfortably on their remote Scottish island, it is already a reality and without this possibility they would inevitably be unemployed and (geographically) unemployable. Even so, it is necessary to counter the hype with some reality: only half our groups suggested that teleworking would become important for them, though nearly 90 per cent of our individual respondents thought teleworking will be common within fifteen years.

The drawbacks are evident at either end of the communications chain. At one end, the typical home is not as yet well prepared for becoming such a workplace. Working on the dining-room table while the children play around you is really no solution. At the other extreme, the traditional office building offers more than just desk space. Above all, it offers us the contact with our peers which we seem to crave. Almost a half of teleworkers feel isolated from colleagues and office gossip. Thus, homeworking may not reduce organizational costs; it may actually increase them in the short term. Even so, some estimates show that perhaps as many as 10 per cent of the working population in some Western countries are already working from home. With careful planning, over time the costs may reduce. Workers visiting the central office can share desks and equipment (often called 'hotelling'). Perhaps more importantly, in the second stage (described as 'dispersion') the office itself can be split up and moved to regional centres where space costs are lower. In the final stage, diffusion – where the organization may be based on computer conferencing networks – the cost may eventually be no more than traditional solutions. Indeed, some would claim that, with the cost of housing a professional office worker in a traditional building being something like £7,000 ($10,000) a year and productivity of those at home rising by between 20 and 40 per cent, there may ultimately be major savings.

The real benefits to be derived from homeworking will accrue to the individual. Not least, the hours wasted in soul-destroying commuting will almost disappear; and the working environment (given the right investment) will improve, though for most of us the view from the window is more likely to be of suburbia than of distant mountains or forest glades! The good news is that almost two-thirds of teleworkers in one survey (noted by Janine Milne, 1995) reported less stress and three-quarters had more time with their family. These benefits will be such that the pressure which we bring to bear on our employers will eventually overcome the latter's objections; and homeworking will slowly grow to represent, by mid-twenty-first century, say, a significant proportion of the total.

The process will probably be encouraged by government in the form of tax breaks and grants since it will offer a considerably cheaper solution to the increasingly impossible demands placed on the transport infrastructure. At its simplest, it will be cheaper to subsidize us working at home than pay for the roads to bring us into offices.

One drawback for teleworkers in developed countries may be that they find themselves in competition with teleworkers from the Third World. As Pete Engardio, writing in *Business Week* (19 December 1994) points out, 'What makes Third World brain-power so attractive is price. A good computer circuit-board designer in California, for example, can pull down $60,000 to $100,000 a year. Taiwan is glutted with equally qualified engineers earning around $25,000. In India or China, you can get top-level talent, probably with a PhD, for less than $10,000.' He goes on to point out that 'India has the second-largest pool of English-speaking talent in the world, after the US. This includes 100,000 software engineers. . . . Dataquest Inc., the research firm, estimates that there are at least 350,000 information-technology engineers in China.' This is potentially bad news for those of us in the West but very good news for our counterparts in developing countries.

Homeworking will slowly become a widespread reality. Though it will not result in major cost-savings for organizations, once the investment in the home facilities has been made it will bring real benefits to the individual – especially where governments subsidize the process.

Electronic Chaos

A potentially more subversive use of the networks is already emerging in the shape of very active use by groups which wish to act collectively, especially where group members are widely spread in geographical terms. Most of these groups will be interest groups: railway modellers in England sharing their thoughts about the ideal layout with their counterparts in New Zealand, or chess buffs who are willing to play anyone anywhere, for instance. Some will want to co-operate towards a common goal. They might, for example, want jointly to work on different aspects of the archaeology of a particular century, or to participate in the human genome project, or to plan the future of a voluntary organization – all of these across country boundaries. More controversially, they might wish to become powerful pressure groups.

A much less welcome outcome will be the opportunity it offers to those who, for whatever reason, wish actively to subvert society. They now have access to the jungle of electronic networks, covering the globe, within which they can conduct guerrilla warfare. Indeed, their tactics can be much the same as the traditional guerrilla, emerging from the jungle only to make a very brief attack, before they melt back into the surrounding cover. There is one important difference. They no longer need a support structure – 'the water in which fishes swim', to quote the famous phrase – the tacit support of large parts of the community. Without the fear of betrayal, they can remain truly invisible.

These will be no ordinary hackers, though, happy just to cause a little chaos with their inane messages – vandalism just for the sheer hell of it. They will be professionals, bent on demolishing parts of organizations, maybe whole organizations, to make their political point or simply to destroy their opponents. Such electronic terrorists will, along with electronic fraudsters, absorb an ever larger amount of police time – together with that of the emerging data protection industry.

Networks offer an ideal medium for groups to share and organize their joint activities around the world. They will also provide an ideal environment for terrorists to create political impact; and combating this will take an increasing amount of police time.

Consumer Power

Parallel with the political power gradually accruing to us will come increasing commercial power. For most of the past half-century, the leading suppliers to the consumer markets have recognized the importance of the consumer under the compendium 'discipline' of marketing. Unfortunately, with the technology of the time, these suppliers could only conduct their marketing activities in terms of groups of individuals, and in terms of average needs and wants discovered by marketing research based on sample surveys. Most recently, many of them have been forced to abandon even this limited dialogue, to meet the needs of the increasingly powerful retailers who have claimed – with some justification – that they were closer to the customer.

This retailer power will, though, probably only have a lifetime of a couple of decades or so. Database marketing, which requires that the supplier tracks large amounts of data on each individual customer's activities, will introduce dialogue with the consumer – this time on an individual basis. Increasingly, suppliers will hold such details for all of us on massive computer databases. They will tailor their offerings, or at least the way they are promoted, not just to the group average but to our individual needs. This will offer significant competitive advantage to those suppliers who invest in the requisite technology and information. In particular, it may in the short term further reinforce the position of strong retailers since they, with their club schemes, have the most direct access to large numbers of consumers and can spread their investments across far more product categories.

The next stage will go much further to allow us to engage in a (more or less) genuine dialogue with our suppliers. I have to qualify this as 'more or less', because suppliers with millions of consumers will still have to mediate this dialogue through computers, but the advantage will go to those with the most 'human' computer – the one which is easiest and most effective (and friendly) for us to deal with. Individual consumers will negotiate with the individual supplier to agree what is to be supplied and at what price – together with how it is supplied – which may then rapidly undermine retailer power. Of course, the negotiation will rarely be in such depth; we will not find it productive to negotiate our supply of canned beans in this way and Heinz certainly won't! But we may use our computer to find which shop that day offers the best package, and price, to meet our current needs. You would enter the list of items you wished to purchase, specifying those where you wanted a named brand and those where you would accept the lowest priced commodity. The computer would then cost the whole list, for each of your local supermarkets, and tell you which one was cheapest. It might even save you the trip by also ordering the goods to be delivered to your home; most large supermarket groups are now experimenting with the concept.

This dialogue, between many suppliers and many consumers, will clearly have the potential for creating immense complexity. In the first instance, we are likely to sub-contract much of this activity to commercial organizations – consumer clubs, say – which will wield the buying power on our behalf. The (human) 'intelligence' needed to run these activities may be further sub-contracted to organizations and individuals in the Third World – as programming and data entry is already being sourced. You may find, in this way, that your baked beans are being bought on your behalf by an Indian, sitting in a mud-hut but connected to a satellite communications system! Eventually, you are likely to sub-contract much of the work, especially the search for information, to more developed versions of the 'agents' (engines) which are already in use on the Internet. These will reside in your computer, and from there will search the world for your requirements.

The ultimate development will be intelligent agents which do not reside on your own computer, composed of packages of code which will float through the networks around the world, that will search out the information you want. The key differences from earlier, simpler versions are that they will have their own separate existence, within the net but independent of their owners, and they will have the intelligence – albeit in a highly specialized sense – needed to maintain their separate existence and to serve their owners. When the technology has become advanced enough, and we trust it sufficiently, these trader agents may even be given budgets – together with the rules for spending these – to buy the information to send back to us. This may even extend to teams of agents working hierarchically together, even though spread around the world, on one owner's behalf. *The Economist* (14 May 1994) suggests, 'In the long term the most intriguing relationships may not be between agents and masters, but between agents and agents. The more agents there are, the more likely it is that they will deal with other agents. . . . If thousands of agents are doing roughly the same thing for their masters, why not pool resources?' This is likely to be most obvious in terms of the police-agent teams roaming the networks to trap illegal operations, especially illegal trader-agents.

Database marketing, creating a quasi-dialogue with individual customers, will bring major changes to the field of marketing in general, and of consumer marketing in particular. Real consumer power will come when individual suppliers negotiate direct, through the net, with individual consumers.

Most of the dialogue between consumers and suppliers will be sub-contracted to commercial providers, probably using cheap labour in the Third World but increasingly moving to electronic agents on personal computers. Such trader-

agents will operate independently across networks, seeking out information for their owners. Ultimately they will be authorized to buy information on behalf of their owners. They may then be operating in teams – not least those belonging to the police, searching out illegal operations and illegal agents.

Net Arts

One aspect of networking I have not yet covered is that of its use by the artistic community – in the widest sense. The individual, working at home, creating his or her own television documentary using material retrieved from the net, is likely to represent a growing market, as the demand for specific categories of education in general, and for 'edutainment' (education as a pastime, as an amusement, rather than as a necessity) in particular, grows.

Authors will also conduct research and sell their wares through the net, despite the problems of protecting their intellectual property (copyright). Those in the visual arts may operate in much the same way – the future hanging space for works of art may be on the computer terminal screen and Bill Gates' cornering of the electronic rights to much of the world's great art may, once more, be a very astute move! As yet, though, the various forms of art have barely responded to the advent of the computer. What is currently being applied to computer graphics, for instance, follows all the old traditions. I suspect that we can forecast that 'something new' will emerge which makes powerful use of the new computer techniques, but it is too early to say exactly what; perhaps the new solid prototyping techniques will revolutionize sculpture and make it more affordable for buyers.

What other impacts it may have on the arts is, as yet, difficult to predict. Anthony Smith, President of Magdalen College, Oxford, states (1993) that 'During the 20th century the conception of entertainment has changed from a number of activities interposed at the edges of the working day to a cluster of industries which attempts to provide for all the leisure time available.' Maybe the twenty-first century will see entertainment merge with work itself. People already talk about it merging with education, to offer a combined package of 'edutainment'. This will pose fascinating new challenges for the process of work design.

Computer networking may also be used – to great effect – by documentary makers, by authors and by artists, with considerable benefits for society as a whole.

Social Highlights

It should be obvious by now that this part of the book contains a number of major uncertainties. While the progressive move away from group loyalties towards individualism is widely recognized as an existing trend, it is not clear exactly where this will eventually take us. Individualism has been described by a number of commentators, but typically in its currently most prevalent form of hedonism. This superficial characteristic is unlikely to remain its most important aspect; the core concept seems likely to evolve into a search for personal fulfilment. What the majority will consider fulfilling, in this context, is yet another unknown – though it seems likely that it will derive, in one form or another, from the range of post-modern lifestyles. It will almost certainly demand fewer rather than more *physical* resources. Whatever the final outcome, the empowerment of the individual will have major impacts on society as much as on the individual.

Patterns of living will change, to take account of the portfolios of lifestyles individuals will adopt, not least in terms of the different patterns adopted to match the lifestages the individuals are currently going through. Education will also change dramatically to meet the ongoing needs of individuals as they change, and as society changes around them.

The resulting changes in society will be just as dramatic. The organization of work will change, though the seemingly insoluble problems of unemployment may soon disappear since they probably are the symptoms resulting from 'revolutionary pains' rather than the structural forces many now imagine them to be. For this to come about, however, it needs to be recognized that they are susceptible to political decisions rather than free market forces. The underclasses may pose a more intractable problem, though even this is likely to be soluble outside the US.

A much greater uncertainty surrounds what form(s) the new community will take and what will result from the progressive breakdown of the family. It is easy to see that there could be many (virtual) communities available to offer the individual intellectual support. What will replace the traditional emotional and psychological support of the previous institutions (the community and the family) is again much less obvious. Certainly, there will be many new attempts to meet these needs, from deploying communitarian values to revisiting the experimental communities of the 1960s, but new institutions will emerge – the need for such support is just too important to consign it to a vacuum. This is one area where we will all be engaged on a gigantic experiment to discover the community of the future. More importantly, we will need to recognise, and accept the validity of, the many new value sets which are emerging with the changing conditions in society.

8

A SHARED WORLD

The third part of this book looks at how changes will affect the world as a whole. In particular, it looks at how the 'unlimited' resources may be shared more equitably, to the benefit of all and the greater safety of the whole of humanity. Like the chapters which have gone before, it is generally optimistic; and like the chapters on technology, it describes an incremental future – the trends are already there to be seen. As a result, there is more certainty in terms of the outcomes, although most of the politicians in the West are as yet blind to these certainties.

Development across the globe as a whole, not just in the West, is proceeding at a rapid pace but in ways which have not been widely predicted. For most commentators, globalization has been seen as an important process applying only to developed countries, and most particularly to the 'Triad' – Kenichae Ohmae's categorization (1995) of the US, Japan and Europe. Indeed, most futurists seem to have been seduced by the excitement of the technological developments in the First World. If they do comment on the Third World, and few do, it is usually to commiserate with its inhabitants on their dismal circumstances and the lack of any positive future for them. In this respect, at least, these futurists could not be more wrong!

The difference which the reports in the media are missing, in terms of developments in the twenty-first century, is that the process will extend to the more than 80 per cent of the world's population, currently in the Third World, which has previously been excluded. The majority of humankind, who are at present barred from the benefits available to those in the First World, will catch up rapidly. More importantly, by sheer numbers they will come to dominate economic and political developments. The prime driver, in this case, will simply be the size of their populations, where one man one vote now carries as much weight in terms of consumption as in politics. This seems to have been overlooked by most governments: until recently the EU did not even have a policy for Africa, the piece of the Third World effectively on its own doorstep.

Above all it will come to represent the ending of the hegemony of the Western nations – the very opposite of most predictions. In reducing inequality and vastly enriching many nations, it will improve the lot of everyone around the world, including those in the Western nations. The gains in international trade will further enrich even those who might otherwise fear it will be to their disadvantage.

As a result of these changes around the world, and impacts of the IT and Communications Revolutions closer to home, the multinationals – the main beneficiaries of the earlier stages of the process of globalization – will no longer continue to make gains in the way that most pundits are also forecasting. The economies of scale, even those applying to finance and marketing which have favoured these multinationals in recent years, may no longer apply in the same way. The corporate planners in the multinationals who have examined the problem have come to the conclusion that, in the information society, there will be little or no advantage accruing to size.

The bases for national advantage will also shift, offering significant opportunities to the emerging nations, which are building their developments on 'greenfield' sites and posing major challenges to developed countries and the organizations operating out of them. During the transition to the new order, before the stability it will offer finally arrives, these processes will sometimes be confusing and painful, not least for those in the developed world – but the final outcome will benefit all, including those seeming to lose in the short term. The global society of the 21st century will, at last, include those currently in the Third World and – within three decades – will be dominated by them.

Demographics

The basics of overall population growth, which will drive these changes, are well recognized and feared by most politicians and media commentators. According to Geoffrey Lean (1994), the world's population (standing at nearly six billion in the mid-1990s) is likely to reach between eight and twelve billion (most probably ten billion) by 2025.

The unspoken corollary, which has been around for some 200 years since Thomas Malthus published his *Essay on the Principle of Population*, is that starvation will soon sweep the world; and, indeed, just over half our individuals expect famine to occur (by 2030) after crops fail (before 2020) – and understandably they rate it (at 8.0) one of the most important issues. The counter-argument is, of course, that it simply hasn't happened. Even so, the latest projections can look alarming; and the fastest population growth is occurring in the Third World, with dramatic impacts where it can be least sustained.

It is for this reason that there has been so much emphasis on population control in general, and on contraception in particular – to the great consternation of the Vatican! Three-quarters of our individual respondents expect worldwide population control to be in place by 2030, placing an importance level of 7.7 on this. In fact – again according to Geoffrey Lean (1994) – the worldwide rate of population increase has been falling since the early 1970s, from more than 2 per cent annually to about 1.5 per cent. This is in part due to contraception, where he reports, 'More than half of Third World women of childbearing age now use it, up from a fifth in 1960' – a little known success story which surprises most Westerners when they learn of it. But, he continues, 'contraception only comes into its own when parents cease to want big families. In poor societies there are good reasons for lots of children. They are economic assets, doing useful work for the family from the age of six or seven.... They provide security in old age.' Geoffrey Lean (1994) highlights the problem with an anecdote: 'When early campaigners in India put up posters contrasting a happy family with two children with a miserable one with six, they found that people came flocking to them to discover how to become like the larger family.'

It is necessary, therefore, for contraception to go hand in hand with development and education (especially of women).

Resource Growth

Third World food resources may still be capable of more dynamic growth than even the optimists allow for. As one specific example of what might be achieved, Fred Pearce (1992) counters the 'dogma [which] holds that [even] desert margins have a fixed "carrying capacity"' by pointing out that in Nigeria and Kenya 'the opposite has happened. Rapidly increasing populations emerge from these studies as saviours of the landscape.' Robert Paarlberg (1994) explains that, contrary to popular opinion, farmers do look to the longer term. 'Hill farmers in some of the poorest countries in Africa have constructed terracing systems and have maintained those systems for hundreds of years.' One of my most memorable experiences was watching a dam being built, in the hills of East Africa miles from the nearest road, by thousands of villagers, building it literally with their bare hands.

Paarlberg also describes the success of the green revolution: 'By switching to highly responsive seeds, more fertilizer use and expanded irrigation, India was able to double its total wheat production between the years of 1964/65 and 1970/71.' On the other hand, there are now widespread doubts about the sustainability of the green revolution. Francesco Bray (1994) reports, for instance, that

'Monoculture reduces diversity. Large-scale mechanized operations hasten soil erosion.' Poverty is, to a great extent, the result of political decisions.

In any case, mass food production may in future become a factory process, with genetically engineered cells grown in vats – perhaps on artificial islands making use of the three-quarters of the Earth's surface (the oceans) currently underused. The limits to growth would then be greatly extended. Growth of the basic food resources in the Third World may not be as limited as some pessimists would predict.

Peace

It may not be obvious from the newspaper headlines, but the world as a whole is a significantly more peaceful place than it has been for almost a century. What is more, this peace is generally the result of civilized behaviour between equals, between nation-states which recognize each other's rights, rather than one imposed by the superpowers as the long peace during the nineteenth century was.

The future of humankind is not, however, automatically guaranteed. There are still thousands of nuclear warheads held ready for use or in arsenals and these all still represent an omnipresent threat, especially those in the increasingly unstable constituent parts of Russia and, almost as dangerously, those in the United States. It only requires one finger to stray and humankind might be thrown back to the dark ages. Our participants worried, in particular, about the proliferation of nations with nuclear capacity. More than two-thirds of our individual respondents believe that there will be risky nuclear proliferation before 2020 (and, understandably, rate this development at the high, 8.0, level). Former Prime Minister of Singapore Lee Kuan Yew even suggests that Japan may eventually abandon its opposition and join the nuclear club.

On the other hand, the nuclear threat, which has hung over the world since the middle of the twentieth century, is now much reduced. Not least, the break-up of the old Soviet bloc has inserted an effectively neutral buffer zone across the middle of Europe, removing the traditional trigger points for conflict between NATO and Russia. Not only that, but *global* conflict is now also almost totally absent; of the recent wars, only a handful have even been between nations; most have been *civil* wars within nations.

The end of the twentieth century and the first decades of the next will be a period of transition, with all the accompanying uncertainties. Global peace has arrived, and a peace dividend is expected, but all the military machines are still in place and it would be unrealistic to expect them to be dismantled in just a few months. The problems will be resolved over a decade or so, but, until that happens, tensions will emerge from time to time as the old strategies and deployments fail to match the new realities.

In global terms, at least, the world is now more peaceful than it has been for a century, and much safer from nuclear destruction than it ever has been since the dawn of the atomic age. Global peace has caught the military–industrial complexes of the major nations on the hop. Their strategies are no longer viable. This problem will, over a decade or so, eventually be resolved; but in the meantime, the mismatches will occasionally result in tensions.

Local Difficulties

In the short term the number of small wars, within countries not between them, has escalated. This is probably due to the prospect of freedom which has been dangled in front of them by the spectacular collapse of the USSR, but mainly resulting from the release of the pressures, on supposedly independent nations, which were previously applied by the superpowers as an integral part of the Cold War. When the end of the Cold War removed the need to underwrite the stability of their client governments, the position of the latter became much less tenable.

Global Guerrillas

The technologies now available to dissidents are awesome in potential, including nuclear weapons and germ warfare. Cetron and Davies predict that, 'the next 15 years may well be the age of superterrorism, when they gain access to weapons of mass destruction and show more willingness to use them. Tomorrow's most dangerous terrorists . . . will not [want] political control but the utter destruction of their chosen enemies.' This prediction was made before the nerve gas attack in Japan showed just how easy it was to acquire such weapons. Perhaps the most important, and frightening, aspect is that described by Davidson and Rees-Mogg: 'A terrorist is unlikely to be deterred from employing weapons of mass destruction by the threat of massive retaliation.' It was such a counter-threat which maintained the balance, and the peace, during the time of the Cold War; but this logic does not impress the isolated terrorist to the same extent.

Globalization will enable dissident groups, including terrorists and guerrillas, to gain support, and operational resources, almost anywhere on the globe. This will make such groups harder to track and control; and their actions may become that much more theatrical and correspondingly destructive of human life. As a reaction, an effective global anti-terrorist police network is likely to be created within the next decade; and the power this obtains may cause disquiet in some more liberal quarters, perhaps even in turn reinforcing the legitimacy of the terrorist groups.

Destabilization

From the other end of the spectrum, the Cold War saw the emergence of large-scale covert actions by the superpowers, especially the US, aimed at supporting small groups opposing 'hostile' governments, or undermining those opposed to 'friendly' ones and especially in destabilizing countries which might give succour to the latter groups. In particular, guerrilla groups were extensively supported, often with the aim of destroying a country's economy rather than gaining the outright victory which they were not in a position to achieve. Angola and Mozambique had their economies destroyed by South Africa with covert US help, although these countries themselves in turn became pawns of the USSR. Lebanon was destroyed by Israel and Syria, each sustained by a different superpower, as was Afghanistan by the USSR and Pakistan (once more with the help of the CIA). Perhaps most cynically of all, Somalia accepted arms from the USSR for its invasion of Ethiopia, only to find that the USSR then armed Ethiopia itself – so Somalia turned to the US for its next tranche of weapons. The majority of these destabilization campaigns have now ended, but their effects linger on. When, over several decades, you have persuaded factions within a country to hate each other – and have then given them the arms to turn that hate into action – it is naive to expect the situation to stabilize as soon as you change your mind.

Many of the instabilities among the smaller nations in the Third World are the legacy of covert destabilization operations by the superpowers. The scale of such operations has now significantly reduced, but their legacy of inherent instability may run for a generation or more.

Global (US) Culture

The main practical impact of globalism (as opposed to globalization) has so far been on culture. In practice, while the emerging world culture borrows elements from a wide variety of national cultures, it is above all an American culture and, especially, a Californian one. As a result, half the world wants to live in downtown Los Angeles or, at the very least, to imitate its supposed lifestyles. Those of us who have actually visited LA will appreciate the delicious irony of this; LA is, in reality, one of the most disagreeable cities in the world.

Needless to say, this is not the image conveyed by the Hollywood studios. With satellite television now girdling the world – you see receiving dishes in even the remotest Nepalese villages – its voice has become the major influence in the global integration of culture which is already under way. This basic message is reinforced by the actions of the multinationals, whose products and services typically incorporate the same messages. Coca-Cola and McDonalds, among others, have brought the American dream to nations around the world. Little of this is deliberate, making the process that much more insidious and potent! The brainwashing is not merely voluntary, but is actively demanded by the victims.

On the positive side, people around the world are now beginning to share a common view of the coming world culture, and this knowledge helps to reduce potential conflicts between competing systems. Unfortunately, the process is still essentially colonial: US culture is replacing indigenous ones in exactly the same way as the British Empire forced its ways on its colonies. The richness of the national cultures is lost; further, the model contained in this US culture is also now outdated, based as it is on the family structures and consumer culture values of the 1950s.

The social changes now under way call for totally different social structures in general, and family structures in particular. There is no longer just one perfect way of life, one happy family. Instead, there are many competing lifestyles, each of which may be just as valid for those individuals who adopt it: the antithesis of the Hollywood nuclear family myth.

Localism

An *Economist* (30 July 1994) editorial records the widespread view that: 'The cliché of the information age is that instantaneous global telecommunications, television and computer networks, will soon overthrow the ancient tyrannies of time and space'. It goes on to point out 'geography still matters', and to say 'such developments have made hardly a dent in the way people think and feel about things. Look, for example, at newspapers or news broadcasts anywhere on earth, and you find them overwhelmingly dominated by stories about what is going on in the vicinity of their place of publication.' Certainly, this reflects my own experience. On my travels, I have often tuned into the local television news and, especially in the Third World, have been astounded by the parochial nature of the coverage, with a concentration on what seemed to be the minutiae of local events to the exclusion of global ones. Returning home, I realized that the BBC was just as parochial, in terms of its domestic channels at least.

The Economist goes on to make yet another telling point: 'even the newest industries are [still] obeying the old rule of geographical concentration ... all but one of the top 20 American carpetmakers are located in or near the town of Dalton, Georgia; and, before 1930, the American tyre industry consisted almost entirely of the 100 or so firms carrying on that business in Akron, Ohio. This is why the world got Silicon Valley in a short and narrow strip in California in the 1960s. It is also why tradable services stay surprisingly concentrated: futures trading (in Chicago), insurance (Hartford, Connecticut), movies (Los Angeles) and currency trading (London). Making general a point about electronic communications, they add, 'The most advanced use so far of the Internet has not been to found a global village but to strengthen the local business and social ties among people and companies in the heart of Silicon Valley.'

A global culture has already emerged, based upon the mythical US–Hollywood model of the perfect nuclear family (living in an aggressively consumerist culture). This culture is outdated in terms of the social changes now taking place. It is, however, the only model promoted and accepted worldwide.

Fundamentalism

Perhaps the greatest challenge, in the medium term, to the global culture of Hollywood comes from religious fundamentalism in general, and from Islam in particular. Adopting a rather more positive viewpoint, it is possible to see that these various religions might offer positive ethical contributions to a combined culture, whereas the greatest weakness of the Hollywood version is its grounding in the vacuous ethics of the consumer culture at its worst. The Islamic approach to business, with its emphasis on the rights of the community, may for instance provide an antidote to the pure worship of Mammon; especially where we are moving to an age of co-operation. In this very positive context, fundamentalism could be seen, at one level, to be a valuable antidote to the trivialism and unreality of the Hollywood culture.

As yet, much of fundamentalism, especially of the Islamic variety, does not just seek to modify the overall culture but to overthrow it. In addition, for the mass of its followers as opposed to its (often fanatical) religious leaders, it is frequently motivated by a reaction against Western (US) imperialism. Thus, it often offers a more powerful, and more rigorous, basis for nationalism than mere patriotism at a time when national patriotism (as opposed to ethnic or tribal versions, which offer another focus) is on the wane. It is, literally, a God-sent opportunity for some emerging nations which have no other easy alternatives. It is paradoxical that the country which is often seen to be the greatest threat, Iran, has a modern constitution, not one based totally in Islamic readings. In fact, it is more likely to come from Algeria – largely ignored by the Western media (outside France) – though it should be noted that it is one of the few countries where the underlying struggle is one between classes: the underclasses, an oppressed majority, are Islamic!

The Islamic religion delivers a very strong core message, powerfully delivered as a certainty – the very word of God written down by Mohammad which has been reinforced in many of its adherents by centuries of repression, typically by the West. Indeed, its growing political power may come from the fact that it is the only major ideology remaining intact (and vibrant), and certainly the only one to champion the rights of the oppressed in the regions from which its power derives. But Islam should not be seen as the new enemy.

Indeed, the senior managers of multinationals operating in the Islamic countries have expressly stated to us that it is definitely not a threat. They warn that we in the West are creating false enemies which then can become real ones.

Like any other modern religion, which much of Islam is struggling to become, there are many different approaches within its boundaries, and the West ignores these differences at its peril. The one major danger which Brian Beedham (1993) sees is that 'most Muslims are still willing to leave interpretation to the little band of self-appointed experts.' It is the bleak, religious conservatism of this small group of isolated (male) religious fanatics to which the secular West objects. Its own similar fanatics of previous times – remember the Salem witch trials – have long since been forgotten. On the positive side of the equation, Islamic economists would point to the moral superiority of concepts such as *zakat*, the word for charity by which rich men improve themselves by giving to the poor. This economic morality also extends to the charging of interest, which is banned. The result is that Muslim bankers have been forced to develop devices which get round this ban. One solution, known as *madaraba*, is significantly different from practices in the West. It requires the bank in effect to buy shares in the company borrowing money. If it performs well, the bank gets an agreed share of profits; but if it fails, the bank also loses. The Islamic approach formalizes the shared risk. The most important practical point is that this is a principle which applies to all such loans. Japanese banks have long had cosy relations with their favoured customers, often claimed to be one reason for Japan's success, but these arrangements are voluntary while *madaraba* is, in effect, compulsory.

Christian Fundamentalists

Putting another spin on the issue, Henry Louis Gates (1993) suggests that 'growth of Islam will fortify the side of faith – the side with which many cultural conservatives in the West have allied themselves. In this sense, the Muslim from South Asia or the Maghreb has more in common with his God-fearing Christian opposite from the Home Counties than with his secular counterpart.' In the short term, it may be the Western Christian fundamentalist sects which pose the real threat of anarchy, on a very local scale, as they adopt the Millennium as the likely date for Armageddon. Ted Daniels, editor of the *Millennial Prophecy Report* and quoted by Simon Hatterstone in *The Guardian* (8 April 1995), estimates that 'there are around 350 American organizations that predict the millennium will bring some form of Armageddon'. More alarmingly, some of them, such as the Sons of God, seem even to think it is their God-given role to create Armageddon themselves! A Newsweek survey in 1994 reported that 61 per cent of Americans still believe in the literal truth of the Bible that Jesus will return to Earth. In light of the increasing instability of the US as a whole, this view cannot be discounted as an indirect force on the future.

In the political context, fundamentalism, especially that espoused by some fanatical Islamic groups, can perhaps best be seen as a reaction against the inequalities arising from Western imperialism. It offers more power, and righteousness, than mere nationalism. It may spread, as the reaction against such inequalities grows, unless the developed nations address the disease of inequality rather than the symptoms of the *jihad* it provokes. It might be possible to obtain the best of both worlds, Islamic and Western, especially in terms of financial activities, if we do not overreact to the imaginary threats.

Global Diffusion

A rather different aspect of global developments is demonstrated in the dramatically increased speed of diffusion of new products and of ideas. Knowledge of an important product innovation may travel around the world in a matter of weeks and copies can be on the market within a month or so, crudely pirated or, with more sophistication, reverse engineered to avoid the new laws of intellectual property. Before this reduction in timescales, product developers could test and debug a product in their home market and then gradually introduce it worldwide. Now they have to launch immediately on a global basis if they are to gain the full profit from their innovation. Such global launches are *very* expensive and the risks are high, as Intel found with a bug in its initial Pentium chip which cost it hundreds of millions of dollars to rectify around the world. The international, multinational players now have to be very well heeled. They increasingly protect their investment by branding ('Intel inside', for instance), which is more difficult to bypass, rather than by pure innovation.

Perhaps even more important is the speed of diffusion of *ideas*. If changing the overall global culture is problematic, individual ideas can develop a global potency almost overnight. Thus, the agents of social change, even revolutionaries, can rapidly learn from developments on the other side of the world. Where Marxist ideas took decades to disseminate, similar developments can now hit the headlines in as many hours. As one example, terrorist tactics may now be global – though, fortunately, as yet few terrorist organizations are.

At a more universal level, people watch, live on television, the emergence of new philosophies wherever they are first propounded. Possibly the most influential television pictures, delivered live to hundreds of millions of viewers around the world, were those showing the destruction of the Berlin Wall, which took place before even the political leaders in the Kremlin realized it was happening! This delivered the lesson that even the most authoritarian governments can be defied and overthrown, a lesson which is now being put into practice in many parts of the world outside the former communist bloc.

In this way, new ideas, new social concepts and political philosophies, diffuse very rapidly without regard for physical distance, or even the social circumstances where they arise or are received. In this sense, the global village is now open to all ideas – only their intrinsic power as ideas now matters. In this vein, it was reported by some of the participants in the Rio summit on the environment, with whom we have since talked, that Greenpeace carried more weight than most nations!

Diffusion of new products and, especially, new ideas has now become much more rapid – in effect instantaneous. As a result, inequalities around the world may be ironed out much faster than expected. Powerful new ideas may now develop their global potency within a very short time, in weeks or months rather than in decades and without regard to distance. This may lead to greater cross-border alliances on single issues.

Global Economics

So far, I have had remarkably little to say about the economic forces which are traditionally seen as having major impacts on the future of humanity. Most of those taking part in our research tended, to our surprise, to downgrade the importance of economics, in whatever form, as a primary driver; and half our individuals even thought there would be a change from market forces as the major economic driver. The groups, indeed, indicated that – as a discipline – it had become too far removed from the real events. Their comments were largely limited to three issues: economic collapse, tariffs and taxes, and economic imbalance between the trading blocs.

Perhaps it is a reflection of the dramatic degree to which economists themselves have now downgraded the power which they claim to hold over the future of even economic developments. When Keynesianism dominated economic theory, it was generally accepted that governments (advised by wise economists) could intervene to control their national economies and, indeed, they did so successfully for several decades. The immediate replacement, Monetarism, soon fell out of favour, to be replaced, after a flirtation with exchange rates (cut short by the ERM fiasco), with a simple reliance on 'market forces' and a simple objective of low inflation rates. Since Keynesianism, the main economic argument has been to the effect that governments could not intervene successfully.

The most fundamental problem may be that traditional economics is irretrievably tied to the age of 'physical' production. Its basically simple 'laws', developed soon after the first industrial revolution, understandably revolve around the exchange of tangible products, classically described by Adam Smith in the context of a factory manufacturing pins. It has recently faced immense difficulties in adapting these laws to the different, often more complex, behaviour of the intangible services which are now coming to dominate trade. Since economics is typically concerned with the rational price mechanisms for sharing out the diminishing quantities of scarce physical resources, the resulting theories offer few useful insights into an economy which is, as we have seen, now based upon effectively unlimited (and, in economic terms, unmeasurable) resources which are quintessentially intangible, bought by consumers who are often less than economically rational in their decision-making.

As a trained economist myself, and one who has long been fascinated by the intellectual challenges the subject offers, it grieves me to admit that the whole discipline may be so closely linked to the previous era of modernism (and, in particular, to Fordism) that it may be in danger of being superseded, in the new age of post-modernism, by disciplines such as sociology and marketing, which offer much more practically useful concepts for describing the behaviour of society.

Whatever the reasons, it is certainly true that economics is now often seen as an essentially theoretical discipline rather than a practical one. This fact highlights the problems caused by government economists' recent focus on the rate of inflation to the exclusion of almost all other economic measures. There is no obvious reason why economic health and low rates of inflation should necessarily be linked in this way. From 1950 to 1973, consumer prices rose by an average of 4.1 per cent a year in the sixteen leading OECD nations, and yet this was the most successful period, in economic terms, since World War II. From 1973 to 1979 they did unfortunately average 9.5 per cent – setting the scene for the subsequent obsession – but this should not justify the level of obsession.

Keynesianism and Bretton Woods

Keynesianism worked remarkably well in its time, even coping with inflation, and the deployment of the theories of demand management may well have led to the long-lived post-war boom, popularly believed to have been undermined by the advent of Monetarism. An equally important but less reported factor was the breakdown, slightly earlier in 1970, of the 1944 Bretton Woods agreement, which may have led to the developments which destroyed the Keynesian consensus.

The original agreement had been one of the great economic landmarks. It created the World Bank and the IMF, both of whose roles changed significantly, however, after the débâcle in the 1970s. Recently, both have functioned as though unrelated to the UN (whose Secretary General is no longer even invited to their annual meetings!). Above all, it created a stable system of fixed exchange rates, which worked well for more than two decades. The destruction of these was probably Richard Nixon's most deplorable, if also least known, legacy! Once the global financial markets had subsequently emerged, Keynesianism on the national scale became impossible.

In the early 1970s there occurred a watershed which separated the optimistic post-war boom years from the pessimistic end-of-century uncertainties. In addition to the breakdown of the Bretton Woods agreements, there was the oil-price shock of 1973. The massive redistribution of currency flows associated with this, together with the effective deregulation of the global financial markets, certainly destabilized governments' economic policies around the world. Energy conservation became the byword and demand fell – and prices returned to reasonable levels. The psychological impact of all these changes was much greater and still remains with us. The doom and gloom surrounding the Millennium is a pale imitation of that which followed that oil-price shock. For many, especially government economists, it was the end of the world as they knew it. It is easy, now, to laugh off the views of these alarmists. Yet it really was the end of the world as we *then* knew it.

In the context of the 'theory' behind this book, at that time the expectations slowly changed from optimism to pessimism. Paul Krugman (1994b) says, 'It wasn't until 1978 or 1979 that the public began to develop a really deep sense of unease about the economic future . . . and the psychological change in the end reached deeper than the numbers themselves can convey.' The beginning of the age of psychological doom and gloom was bad enough, but, as Krugman adds, 'The 1970s saw an astonishing rise in the influence of strongly conservative ideas in economics (as well as in other areas), a rise that was only certified by the 1980 victory of Ronald Reagan.' Is it surprising that a conservative viewpoint (a return to safe traditional values) came to the fore?

Economics has declined in importance as a major practical force determining the future of humankind, to be replaced, perhaps in the short term only, by market forces, which are not susceptible to government intervention.

The death-knell for Keynesianism probably came with the demise of the Bretton Wood agreement, which led to the financial markets' destabilization of any such national planning.

The Global Casino

Despite the problems which emerged after the end of the Bretton Woods system, until quite recently the financial markets were claimed by Western governments to be the embodiment of all that makes free enterprise such an all-conquering ideology. But now the markets themselves have moved on, and the prevailing myth today hides another great and growing industry – that of professional gambling on a global scale. The pensions of teachers are wagered alongside the earnings of the multinationals. In the UK, at the most basic level they no longer operate as the major source of external capital for new ventures. That role – for the relatively small proportion of corporations which do not now fund all new developments from their existing cash flows – has been largely assumed by the banks. Nor do the markets now generally seek to act as a lever for shareholders to maintain control over their wayward charges. That role has been largely subsumed by the legislators. The scale of operations which the new technology allows has meant that the resulting surges of money, microsecond by microsecond, backwards and forwards across three continents, have come to dominate all money flows. Even in the global money markets, the speculative capital flows have dominated the operational flows of money needed to finance international trade. Bezanson and Mendes (1995) reported that 'over US $1 trillion ($10^{12}$) worth of foreign currencies changes hands each day', adding, with a degree of understatement, that this leads 'to concern in several quarters, including central banks, over market volatility and even meltdown'.

A second significant development was the new-found freedom to arbitrage over time, now dignified, as a group of processes, with the term 'derivatives'. The importance of the market was no longer to be where it currently stood – a rational, measured reflection of the net worth, for example, of quoted stocks – but in where it will be in just a few microseconds. The blinkered focus is, in this way, always on trends and never on what underlies them. All is relative, and nothing is absolute.

The nature of the new markets is, therefore, of vast capital flows surging towards perceived future changes, no longer constrained by any need to reflect actual events. The analogy of the casino is justified: the obsessive preoccupation of the participants is with predicting how the other players will place their future bets. The obvious downside is the possibility of a global financial crash – expected, by two-thirds of individuals, to occur before 2020.

Trading Blocs

Ever-larger trading blocs are emerging, including the EU and NAFTA, which potentially have the power to oppose global financial markets on some issues at least. Javetski and Glasgall predict that: 'Supercurrencies spanning entire regions may emerge, dwarfing Western Europe's recent quest for one unit of exchange,' distorting some aspects of world trade, especially in terms of exports and imports between small and medium-sized enterprises.

Japan is recruiting Third World countries on the Pacific Rim into its own trading bloc. It also seems inevitable that NAFTA will similarly recruit the countries in South America, starting with the members of the Mercosur bloc; this is a logical extension of the Monroe Doctrine. This only leaves Africa, which might seem an obvious target for the EU but which is, as yet, barely mentioned in EU foreign policy. The European Commission is now reviewing its position, but even now seems to find the Mercosur bloc more attractive!

In earlier times capital flows reflected the real, operational flows; today the real flows must, in some distorted way, mirror the imaginary. The global financial markets are now dominated by speculation – making them by far the largest casinos, undermining their traditional functions and leading to the destabilization of many economic activities. It seems likely that the world economy will be split into three main trading blocs, each comprising major trading nations together with Third World clients. These blocs will be Japan with the Pacific Rim, NAFTA with South America, and the EU with Africa.

Redistribution of Wealth

I have already stated, at some length, that global wealth, overall, is sufficient to achieve almost anything we might desire. The problem is the distribution of that overall wealth, especially to the underclasses and to the Third World. The much reported fear that development of Third World economies will reduce living standards in the developed world can be largely discounted, despite the politicians' rhetoric. As Paul Krugman – Professor of Economics at Stanford University in California – explains: 'Fears about the economic impact of the Third World are almost entirely unjustified. Economic growth in low-wage nations is in principle as likely to raise as to lower per capita income in high-wage countries; the actual effects have been negligible.... When world productivity rises (as it does when Third World countries converge on First World productivity), average world living standards must rise; after all, the extra output must go somewhere.' But Pam Woodall (1994) suggests, 'Within each country there will be losers. In particular, as increased competition from low-wage economies reinforces the impact of changing technology, many unskilled workers will lose their jobs or see a drop in real wages.'

Economic policy is then likely to focus, in the form of socio-economic policies, on how redistribution may best be effected, to the benefit of all including the already rich! These new models are likely to revolve around decisions, on a global scale, which may prove to be the forerunners of global economic management which may, in turn, be the stepping stone to full global government – though only just over half our individuals expected even a single global economy to emerge, and then not before 2030. In the shorter term, a single global currency is likely to emerge – predicted by more than half our groups – which will offer some stability within which the global changes may take place.

Management Focus

Joseph Coates, admittedly coming at the issue from the direction of forecasting the future of science and echoing what many econometricians believe, could not be more wrong when he suggests (1994): 'Applied economics will lead to a greater dependency on mathematical models embodied in computers. These models will have expanded capabilities and will routinely integrate environmental and quality-of-life factors into economic calculations.' In reality, the one thing econometric models have in common, apart from holding a fascination for financial journalists, is a propensity for failure. Macro-economic theories are coming to be replaced, in practice, with socio-economic ones, while micro-economic theories have been largely superseded, again in practice, by management theory.

Corporate strategy theory, particularly the investment in people, (best encapsulated as Human Resource Strategies) has recently come to the fore, stimulated by the success of the Japanese corporations which have pioneered it. It is true that there was a temporary rejection of 'lifetime employment' by Western companies (though not by Japanese ones) when the recession at the beginning of the 1990s forced them to concentrate on short-term cost savings. Indeed, there were wholesale reductions in their workforces. We still believe stable employment of this kind (for the parts of the employee's life when he or she chooses this) will be one of the mainstays of future policy. Wherever skills are scarce, as they will continue to be, it makes sense to husband them. Fortunately, most management theory has, so far, been eminently sensible as well as pragmatic and flexible enough to help in practice. As a result, the emphasis on management theory rather than economics will probably continue for the next couple of decades.

The main change, at the end of this time, may be a shift to the social issues involved in supporting the individual rather than, as now, the group or organization. Several years ago, we suggested that a new corporate 'accounting' measure, 'Benefit', should be introduced. This would parallel 'Profit' which is internally generated for stockholders. 'Benefit', on the other hand, would be externally generated for all the stakeholders. It would include the net gain (or loss) to the community, and to society as a whole, as well as to the workers and shareholders. In any case, many of the new developments may be in the public sector. As Clive Crook (1993) suggests, 'With greater wealth come greater demands for public goods that unassisted capitalism may fail to supply.'

Organizational Structures

At the same time, the IT and Communications Revolutions are stimulating the introduction of new organizational structures. Cerys when the typical chief executive (CEO) rules a rigid iming responsibility for every success (though not every umbered for most organizations. The structure under- ; transformed. It will also result from the abuses which w perpetrating. Their increasingly obscene salaries – voted to them by subservient boards, often reaching one hundred times that of lower-level workers compared with no more than ten times in Japan – are offensive enough; but the fact that these packages are performance-related, and hence demand measures leading to maximization of very short-term profits which benefit nobody but the CEO, is increasingly recognized as a fundamental flaw in Western business structures.

Some of these new structures, such as the matrix structure (which superimposes, say, a project reporting line on top of a functional one) have been around for several decades. Others, such as self-managed group working, are relatively new. Our contacts among the senior managers in the leading multinationals suggest that the secret now, for governments as much as for businesses, is 'learning how to make small fixes to small problems'. Some structures, such as cellular organic structures, which are hybrids of the previous two (in which the cell, the working group, is so flexible that members can come and go, and reporting lines change, as its role changes), are as yet only being used by a few organizations.

Whatever their exact form, the structures which will dominate in the first half of the twenty-first century are likely to be those which favour relatively independent, peer-to-peer, relationships such as those which currently apply, for example, in consultancies. They may even become collegial at the extreme, as in a faculty of a university, where its members manage themselves. This does not weaken the work discipline; rather, it changes its nature. There will, thus, be a flux of new structures which will be tried out over the next few decades and most will give significantly more power to the individual.

Competition or Co-operation

The traditional Western (capitalist) model of business strategy – or, more widely, of interactions in general between individuals but especially between organizations – has come to be that of competition. It has frequently been developed, in the theory at least, to apply to situations which are described as 'zero-sum' games: those in which the two parties interact on the basis that if one 'wins', the other must 'lose'. The most frequent outcome, paradoxically perhaps, is that they both choose to lose! In the management field, the theory of competitive advantage, making your own organization (or even country) more competitive, has most popularly been developed by Michael Porter. His ideas were behind many of the developments in corporate strategy during the 1980s. They are also, in many respects, the key (but often hidden) concept underpinning the emergence of the 'market' as a major political force. Thus, two products, or ideas, face each other – in a parody of a Western gunfight – and fight for the consumer's purchase decision. The winner takes all, even if the loser does not die immediately.

Our own research shows that this has never, in recent years at least, been what normally happens in practice. It is yet another myth, though a very powerful one. Almost all organizations conduct co-operative relationships with their customers – they would be foolish to do otherwise – and increasingly invest in these all-important relationships. The majority of the Western organizations in our own research even indulge in co-operative relations with their competitors and the European Commission seems to be preparing to reinforce this message of co-operation.

Management theory has largely replaced micro-economics as the practical framework for organizational decisions; and corporate strategies are increasingly driven by Human Resource Strategies. Such management theory will continue to dominate, replaced eventually by a focus on social issues posed by support for the individual.

Organizational structures are likely to be based upon peer-to-peer communications and self-management, giving increased power to the individual.

Most organizational relations are already co-operative rather than competitive, and co-operation will become the accepted model for almost all transactions.

The Myth of Multinational Power

For a long time the most pervasive myth surrounding globalization related to the power supposedly held by the multinational corporations. Indeed, four-fifths of our individuals expected them to dominate business life within two decades. Upon inspection, the coming domination by multinationals turns out not to be true. The largest hundred do control 60 per cent of global assets (worth more than $3 trillion in terms of worldwide assets in 1990). These multinationals, especially the top one hundred, certainly are powerful as individual corporations since they already account for more than half the world trade in manufactured goods and services. In terms of changing the lifestyles of people around the globe, they also wield enormous influence. The basic myth (of multinational hegemony) is, however, quite wrong in that it presupposes that they are increasingly working in concert to become *de facto* the new world government, with all the power and none of the responsibility.

One of the more surprising aspects of our earlier research among the multinationals was the discovery that they had no real sense of being multinationals. They shared some sense of identity with their competitors in the same or similar fields, but they saw no relationship with multinationals in other fields. As a result they could not conceive of working together with other multinationals in general, even if it meant that they could together control the global environment to their advantage. When IBM, at its peak, probably had the power itself (without involving any others) to change the future developments in its own environment, it did not even consider taking such actions. This was not because it was repelled by the prospect, though the constant threat of anti-trust actions would probably have caused it to think twice, but because it never occurred to its management that such actions, shaping the future of society, were the province of a business organization. My other research comes to much the same conclusion, whatever the world outside may fear, Western organizations are very clear that creating the future of society is the responsibility of government not of business – a very different view from that held by the organizations which were embedded in the tradition of Marxist economics. The power of the multinationals, therefore, is limited to that within the boundaries of the individual organizations.

Another myth, perhaps even more pervasive, is that the large firms in general, and the multinationals in particular, are losing ground to small firms. John Naisbitt even made this theme central to his 1994 book *Global Paradox*, a theme which he summed up as, 'The bigger the world economy, the more powerful its smallest players'. The apparent buoyancy of the small firms sector, which seems to lie behind his latest theories, arises from a misunderstanding which Bennett Harrison (1994) describes as 'concentration without centralization'. The changes reflect a concept, described in labour economics as 'core and periphery', where core workers, those who are essential to the business (and in whose training the organization has usually made a significant investment), are well treated – with high salaries and lifetime employment – and the 'peripheral workers', who can easily be replaced on the general labour market, are hired and fired as needed. The latter category are often shifted to sub-contractors and have, in this way, fuelled the rapid growth of small firms.

Taking the long-term view, it is possible to speculate that this may be yet another symptom of revolutionary pains. As more and more workers become ever better trained, and as labour becomes scarcer, the periphery is likely to shrink dramatically. Organizations will have to enlarge their 'cores' to protect their supplies of skilled labour; and 'lifetime' (or at least long-term) employment will probably make a come-back in most large organizations.

9

THE DEATH OF NATIONS

This chapter addresses the new realities of global power and in particular the relative decline of the nation-state, which is steadily losing its powers – downwards to communities and upwards to the new supranational groupings. These resulting shifts in power will have an impact on populations around the world.

To start I will examine the future of the most important individual nations in some more detail. Many other commentators organize their review of future developments nation by nation, arguing that the character of each nation will be the key to its future. As you will by now be aware, I do not generally subscribe to this view; believing that the forces at work are bigger than nations, I generally choose to describe them in global terms. Even so, national characteristics will be, in some key instances, very influential. In particular, the first nation we will examine, the United States of America, can be seen as the exception that proves the rule; its idiosyncratic future is likely to be determined largely by its national characteristics.

United States

Some nation-states will still continue to represent major forces in their own right. Of these, the United States will continue to reign supreme for some time to come, but its development will be atypical. The viewpoint put forward by our groups was that the US, alone of all countries, has probably gone too far along the road to creating a large, ungovernable underclass. It seems to be the only nation where the gulf between these unfortunates and the other, relatively rich, members of society has become institutionalized – creating two nations – in a way which has rarely been seen in developed countries since Victorian times. This schism is already the most characteristic feature of much of US society, and is likely to dominate the future workings of that society. There is a wide range of opinions on the subject. Marvin Cetron (1994), optimistically, sees that, as a result of his forecast of a move to more progressive taxes and benefits, 'The [numbers of] very poor and the very wealthy will decline in American society'; but this very optimistic view of the US was definitely not shared by any of our groups. In the US the rich are increasingly denying the poor any significant share in the considerable wealth of the country. Further, this represents an absolute reduction in the latter's incomes, not just an increase in relative deprivation. As *The Economist* (5 November 1994) points out, 'In both America and Britain, income inequalities are now larger than at any time since the 1930s. But there is one big difference between the two countries: in Britain average real incomes have risen fairly briskly over the past two decades; in America they have risen only slightly. Combined with greater inequality, this means that America's poor, unlike Britain's, have grown poorer absolutely in the past two decades.'

Even working-class Americans have said that large differences in income, and hence an indefinite continuation of the existence of the underclasses, were necessary for their own continued prosperity. While full of fear for the outcome of their actions, the US rich barricade themselves into fortified ghettos. The poor, on the other hand, set up their own crime/drug-funded states within a state. As a whole the US economy is living on borrowed time, as external assets are run down to pay for the profligate lifestyle its citizens and government can no longer afford. Hamish McRae (1994) records that 'The figures for savings show that the US saved less than any other large advanced industrial nation during the second half of the 1980s.' As the Brookings Institute has pointed out, this may not be as much of a problem as it might be for weaker countries since the US can probably continue to run down its assets for a century or more before it is declared 'bankrupt'. Having said that, however, the rest of the world will increasingly look askance at these two aspects of national character: the failure to treat all members of its society fairly, with the consequent emergence of a crime-dominated culture, and the refusal to live within its means. This questionable behaviour will probably lose it the political leadership of the world, even of the developed world, which it currently enjoys.

The United States will probably come to exemplify the apocalyptic alternative for the future. As such, it will offer a useful lesson for the rest of the world, reminding backsliding nations of the dark forces which might also overwhelm them – in a paradoxical reversal of the role in which the US previously cast the communist regimes! In particular, the real fear felt by the mass of its population, in the cities at least, will be conveyed to the rest of the world; and its image, especially on television, will increasingly be created by pictures of its underclass no-go areas burning out of control. This may be an unfair representation of much of the US, of middle-America for instance, but a devastated downtown LA will replace an idealized Beverley Hills as the symbol of the new US.

Another aspect is shown by the number of US lawyers: 756,000 in 1990 (compared with 260,000 in 1960 and fewer than 15,000 in Japan!). Hamish McRae offers the comment that 'there is some evidence that the number of lawyers in a country is in inverse proportion to its growth rate. Too many lawyers, so it would seem, actually destroy wealth.' The result of this growing gulf is that, again according to Hamish McRae, 'Edge cities, and the decline of inner urban areas, are a key feature which will distinguish the US from the rest of the world.' Brave political actions might rescue the situation; but it would have to be a very brave, not to say foolhardy, politician who took these actions. During his first term, Bill

Clinton learned – the hard way – of the impossibility of attempting such radical solutions.

US Economic Power

Even so, the United States is likely to remain an economic superpower, as well as a military one, for the foreseeable future. Despite much of its industry growing uncompetitive on world markets, it will continue to be protected by the almost monopolistic access it has to its own domestic markets. In addition, despite the gloomy prognostications in the previous paragraphs, most US citizens will live in acceptable environments. Indeed, there are anomalies behind the otherwise harsh social philosophies held by individuals there. Peter Drucker, for instance, reports (1993) the surprising fact that 'Every other American adult – 90 million altogether – works at least three hours a week . . . as a volunteer for a non-profit organization.' Above all, its leading-edge, high-tech industries will continue to lead the world. The anarchy, which will be tearing the rest of the US apart, will here – in its purest entrepreneurial form – continue to create the great technological innovations of the first half of the twenty-first century. Its computer corporations will dominate the development of the IT Revolution. Its media corporations will rule the communications world. Even NASA will continue to lead the development of space.

Furthermore, the US will not become, as some fear, irretrievably isolationist; this is no longer a viable option. Despite the gulf between it and the rest of the world, it will not be able to avoid the global forces which are sweeping the world. In any case, it is too far down the line with NAFTA (which will probably expand to cover most of North and South America, and Bill Clinton is already making overtures in this direction); and its dominant transnational corporations could never afford for it to withdraw from international affairs and the coming global society.

The growing schism between the rich and the underclasses will increasingly characterize US society. It, and the run-down of its economic wealth (not least due to the cost of crime), will lose the US the political leadership of the world.

Despite its social problems, the US will remain a powerful economic force; and its high-tech industries will continue to lead the world for the foreseeable future. It will not lapse into isolationism.

Japan

Despite its financial crises during the 1990s, this nation is already a leading superpower; and, in many respects, it is leading the world into the twenty-first century, including facing up to the problems caused by its rapidly ageing population. Its managerial practices offer a model for the rest of the world. It has a very pragmatic blend of government intervention and private-sector entrepreneurship, both geared to the long term. Though MITI (its renowned Ministry of International Trade and Industry) assured us – in our contacts with its senior management – that it now works by encouragement rather than by directive, this amounts to much the same thing in the Japanese culture and has proved to be just as difficult for other nations, immersed in free-market dogma, to copy.

The question which hangs over its future is the strength of its existing culture. How, for instance, is it to take advantage of the moves to individualism which characterize many of the developments elsewhere? Japanese culture has very effectively emphasized membership of the group, to the extent of demanding conformity from individuals. The definitive saying, endlessly repeated to schoolchildren, is: 'The nail which stands proud is hammered flat.' Will the existing culture, which has similarities with others in the Far East, come to dominate that of the other Pacific Rim countries? Evidence suggests that the upcoming generations in Japan are rejecting the older culture – as a matter of principle, not just as the result of the normal generation gap – and wholeheartedly embracing the Western model. This will pose rather different tensions within Japan.

Japan will continue to be a superpower, leading the Pacific Rim countries. It is not clear, however, whether it will impose its existing culture on the region, or will also move to the philosophy of individualism which is emerging elsewhere.

European Union

In the West the most dynamic 'nation' will be the European Union (EU), despite its many setbacks and the dubious future predicted by some commentators. The EU already has the richest internal market in the world and will continue to hold this position for some time. It is also a market which still has considerable growth potential within its borders and, unlike the US, is not in economic retreat. Moreover, it can increase this potential even further by incrementally adding the nations on its borders, especially those to the East. Our individual respondents were especially positive, with three-quarters expecting a 'United States of Europe' (coming into being as early as 2020), compared with less than half believing it might disintegrate (by 2025) or become isolationist (once more by 2020). The real strength of the EU is now emerging from an idea rather than from its resources. Indeed, the EU's new vision is of itself as the model for the world of the future, bringing nations together whilst protecting the rights of their citizens.

The richness of the idea enshrined in the EU, and its importance for the world as a whole, is the real driver. The EU is the first continental-scale federation to be created by the voluntary decisions of its member states, as opposed to a grouping by conquest. There have undeniably been problems with the process, which have been very widely reported – often sensationally so – but these pale into insignificance compared with the scale of its successes. Hundreds of millions of people, in previously independent nations speaking almost as many languages as there are members, have peacefully come together to form first an economic union, a major step in its own right, and then a political one – an unprecedented step. Thus is emerging not just a major new trading bloc but potentially the biggest superpower of all which might ultimately lead the way to true world government. John Rockfellow (1994) suggests that 'national parliaments will be discarded in favour of regional and European bodies, with the regional units becoming the key administrative units.' A majority of our groups (and three-quarters of individuals) also foresaw something like this. In view of the problems involved in relinquishing national identities, this may be a long process; and our individuals, perhaps realistically, expected it to take at least a quarter of a century.

Even so, the power of the idea may be such that the progress of the EU, in terms of population coverage at least, will escalate dramatically. It may well take in Russia, to stretch from the Atlantic to the Pacific the long way, though Jacques Santer seems adamantly opposed to this – even when I pressed him, at a recent (informal) meeting, he would not soften his position. The EU may eventually also capture the emerging nations of Africa which have long had affinities (albeit colonial ones) with individual nations which now belong to the EU. The EU has only recently started to develop a positive policy for developing its relations with Africa. Previously its only official position was 'charitable': overseas aid and support embodied in the various Lomé agreements.

Those taking part in our debates thought it might even ultimately take in India, with colonial links and an inevitable wish to avoid being outmanoeuvred by China in its own immediate sphere of influence – though it just is possible that India will join Indonesia to form a fourth alternative. In this way, the 'EU' may ultimately come to include as much as a third, perhaps even as much as a half, of the world's population – much of it previously parts of the empires of its members – and to set the pattern for much of the rest.

It is arguable that we are actually seeing, in the EU, the birth of global government. Almost everyone (from world leaders, with whom we have argued the point, down to our own groups) is, justifiably, dubious that the existing United Nations organization can ever make the transition to a genuinely viable supranational government. Half our individual respondents even forecast that it will collapse before 2020! The hopes for something global, which were widespread at the time of the Gulf War, have long since disappeared. Indeed, many we have recently talked to, again at all levels (including some world leaders), see the UN eventually going the way of the League of Nations – and in private are actually using those very words! This is a major problem. Even so, half our individuals expected to see a world government formed by 2040. Where the incremental growth of the EU is likely to eventually result in a significant proportion of the world's population being contained within its boundaries, the solution may be to hand.

The inherent power of the EU, with even a quarter of the world's population under its umbrella, will steadily move it towards becoming a *de facto* world government. In this context, it is worth noting that half our individuals saw the EU as encompassing a third of the world by 2025. At that point it is quite possible that the Pacific Rim federation, which will probably follow some way behind in terms of political development (though not as an economic superpower), will also recognize the inescapable logic of joining the putative 'world federation'. The Japanese, at least, are no slouches when it comes to recognizing a good thing and would want to join if only to protect their markets. An India/Indonesia grouping, if it came into being, would probably be even more likely to join such a 'global federation'. Only the NAFTA group, still dominated by the US, probably won't sign up, in the first instance at least.

The greatest virtue of this approach to world government is that it is incremental. At every stage there is a logic for the EU, or whatever it changes its name to next (perhaps just 'The Federation'), to exist in its current form and to progress to the next enlargement. Clearly, the larger it gets, the better it is for all members. That idea has already been accepted, which is why it is so powerful. The rationale for a progressive enlargement of its boundaries, without any obvious limits, has already been accepted by all its existing member states. The European Union is considering, despite Jacques Santer's opposition, the incorporation of Russia – which has now firmly indicated its wish to join. By definition, that would mean that the EU would span more than half the globe. Thus, 'world government' may come about quite naturally, and gently, as the final stage of this growth.

This eventuality may be the most radical suggestion in the whole of this book; and it certainly might be the most important one. The surprising fact is that it has only recently been discussed by even the most visionary members of the European Commission, and not at all by its national states. Perhaps they consider it too revolutionary for publication. The opponents of a federal Europe have long claimed, in the context of defending national sovereignty, that the bureaucrats in Brussels have global ambitions, but it is obvious that even they do not believe this in the literal sense. Perhaps they should, for I hope the argument I have made above shows just how sensible this might ultimately be! This process will, however, be a lengthy one, probably taking a number of decades to reach fruition.

The most dynamic growth area, in the West at least, is the European Union. Its existing market is the largest in the world, and this will grow rapidly as new nations – especially to the East – are added. Its greatest strength is the idea it incorporates; which may ultimately lead to a viable form of world government.

Third World Power

As indicated in the previous chapter, the big change in the distribution of global power, apart from the incremental growth of the EU mentioned above, will be the emergence of the current Third World states. This is basically a matter of demographics. Surprisingly, in view of the low priority given to the subject by most of the media, more than half our groups commented upon aspects of the Third World/LDCs (Less Developed Countries). Three-quarters of individuals predicted that they would achieve 'take-off' by 2025, though only two-fifths expected them to dominate politics (and even then not until 2035). Most of the general groups also specifically stressed the shift of production to Third World countries, as did more than four-fifths of individuals (as early as 2020).

The impact of these new expectations amongst Third World populations should not be underestimated. Pam Woodall, economics editor of *The Economist*, sums up (1994) the economic impacts: 'Over the next 25 years, the world will see the biggest shift in economic strength for more than a century. Today the so-called industrial economies dominate the globe, as they have for the past 150 years or so. Yet within a generation several are likely to be dwarfed by newly emerging economic giants. History suggests, alas, that such shifts in economic power are rarely smooth.' The last comment may be a considerable understatement. As yet, few in the West have even considered that their days as 'top dogs' are numbered, and even fewer – if any – have worked out strategies to deal with this dramatic change in their fortunes. The reality may, therefore, come as a considerable shock to many of its leaders. Pam Woodall details some of the changes: 'By 2020 the rich world's share of global output could shrink to less than two-fifths . . . [and] as many as nine of the top 15 economies will be from today's Third World. Britain might scrape in at 14th place, compared with eighth today.'

The Pacific Rim countries have already made their dash for growth, to emerge as major challengers, in economic terms at least, to the West. Two-thirds of our individual respondents even expect them to *dominate* the world by 2030 – which, if you include China and India (as well as Indonesia) amongst their numbers, they probably will! Our contacts in the multinationals confirm that the South American countries also are some way along the same road, following the model of South-East Asia. This leaves Africa, which is only now starting its own climb out of the abyss and still seems, at times, perilously close to falling back into it. Of its two main potential economic superpowers, South Africa is, at long last, on its way to full economic development. The other, Nigeria, potentially by far the richest on the continent, is, if not quite in a state of civil war, still tearing itself apart economically and politically.

Pam Woodall points out the accelerating pace of development:

'After the industrial revolution took hold in about 1780, Britain needed 58 years to double its real income per head; from 1839 America took 47 years to do the same; starting in 1885, Japan took 34 years; South Korea managed it in 11 years from 1966; and, more recently still, China has done it in less than ten years.' We may thus come to see a relatively prosperous Third World in much less than our lifetime.

Corruption

Unfortunately, the Nigerian dilemma, mentioned earlier, illustrates another set of legacies. Reflecting processes also at work elsewhere, it has been the victim of a form of post-colonial corruption indulged in by Western firms. Having achieved independence, the politicians in such generally very poor countries have too often been subverted by bribes from overseas companies trying to gain control of their country's resources and markets.

This is a problem for a number of Third World governments, which are now being derisively described, by the multinationals dealing most directly with them, as 'kleptocracies'! It is not clear if corruption is on the increase worldwide, although the frequency of media reports to this effect suggests this might be the case. My own suspicion is that the increased level of reports in fact reflects the very opposite trend: the population in general is becoming more intolerant of such corruption and is now complaining about practices it would previously have ignored. If so, that is the one positive sign in a generally bad picture.

Perhaps a slower, staged approach to growth might be less fraught. William Overholt, Managing Director of Bankers Trust in Hong Kong, stresses (1993) that in China, 'Deng's initial farm reforms doubled the income of China's farmers, winning the support of a group that comprises over eight hundred million people – not a bad start as a coalition,' and a power base which is much less susceptible to corruption. The government of Ethiopia, with which I worked, successfully adopted the same approach, first growing the income of the more than 90 per cent of the population who were subsistence farmers. Its subsequent national plan emphasized the same theme; the first priority for the country was 'agriculture-led industry'.

Migration

Much has been made of the potential problems which might be caused by mass migration from Third World countries to developed nations, especially to Europe, from both the East and the South. The US is already being inundated with illegal Mexican immigrants. Our groups considered that such migration might increase to the level where it could cause dramatic problems in a number of regions; and three-quarters of individuals thought that this would be the case as early as 2020. On the other hand, the situation might be defused by the improvements in the home countries of these potential immigrants. John Parker (1995) makes a very relevant point, in his case about migration from the countryside to the cities (another major problem area): 'The main reason why cities are pulling in migrants from the countryside is not the lure of the jackpot, but rural unemployment caused by improvements in agricultural productivity. The migrants would rather have the chance of a job in the city than the certainty of unemployment in the countryside.' It has even been argued that, very much against what the green lobby might predict, cities in the Third World are ecologically efficient: they use far less energy per head and they reduce birth rates. Another child in a farming community is an investment in the future whereas one in the city is a drain on family resources.

The position would be considerably aided if the developed countries increased aid significantly to these countries. Fred Bergsten gives a relevant example: 'The old Federal Republic of Germany realized it could avoid a deluge of East Germans only by paying huge subsidies to induce them to stay at home.' This policy certainly worked, at a horrendous cost, but maybe at a lower cost than the alternative.

The costs may not, in fact, be so high. John Parker, again, adds the footnote that 'most rural migrants . . . wait until they see a chance of finding work, often with the help of family or friends who are already in the city. A study of migrants from Punjab in New Delhi found that 70% had found jobs within one month of arriving, 94% within two months. Anecdotal evidence from Western cities suggests that the same thing is happening there.' If cities in the Third World are so bad, why are so many people clamouring to live in them?

The improved expectations of the Third World will continue to drive the emergence of its nations as the coming economic and political powers. This process is already established in Pacific Rim countries, is well under way in South America, and is now also starting in Africa.

Corruption is rife, especially in the Third World – where it holds back economic development. With an increasing level of public distaste for corrupt practices, we may see bribery and corruption decline.

China and India

As recently as the 1960s, the nations of Asia were usually thought of as 'basket cases' by the West. Yet China and India, which already dominate their local economies, will soon loom large in world affairs by virtue of the sheer size of their respective populations and despite individual purchasing power still being low within each country. The great majority, 85 per cent, of individuals quite reasonably thought that China would become the world's biggest economy by 2030. In Hong Kong, now one of its regions, China already has one of the new 'city states' which are going to dominate elements of international commerce in the future. Deyan Sudjic (1995) makes the point that, by 2005, the Hong Kong conurbations (including Shenzen) will contain forty million people, the largest city in the world. William Overholt (1993) reports that 'China has grown faster economically than any other large economy in history. . . . At the heart lies Guangdong province. . . . Since China's economic reform began in 1979, Guangdong, with Hong Kong management, finance, technology and marketing, has achieved an average annual real growth rate of over 12 per cent.' He further adds, however, Guangdong was to be an airlock through which China dealt with the outside world. Shenzen would be Guangdong's airlock to Hong Kong, and Hong Kong the direct window to the outside world.' Hence, it has to be recognized that the dramatic results recorded by all these regions are a highly leveraged reflection of China's overall policy. Even so, their dramatic successes cannot be ignored.

Since both of the new superstates, India and China, have a history of producing large numbers of hard-working entrepreneurs, previously working their economic magic abroad, their economies will probably be dominated by many small firms, at least in the short term. Their agriculture sectors will remain relatively large, probably protected by government action.

The two countries share many similarities, even in elements of their approaches to their economies; India has often used the Soviet model, and China has turned a blind eye to incipient capitalism in the country. Although India is coming from a colonial, free-market inheritance and China from a Maoist command economy, it seems likely that their separate development paths will eventually converge, depending upon which of the federations they choose to join. In fifty years or so they might look very similar.

All may not be plain sailing, however. William Overholt stresses that: 'More significant than ethnic strains in China are differences between the more prosperous coastal regions [now, such as Guangdong, rapidly growing ever more prosperous] and the poorer interior regions. Historically, frictions between the coast and the interior have been serious, and indeed such frictions played a big role in the convulsions of China's civil war.' Despite the enormous tensions which these changes will engender, it seems possible that transformations in both China and India will be achieved relatively peacefully, but allowing for their vast sizes there may be a number of local difficulties, each of which would destroy a small nation.

As a result of their vast populations, China and India will both become major superpowers, sharing many characteristics of economic development as they move towards similar mixed-market economies from opposite directions – with relatively few major problems.

Regional Spectrum

Perhaps the most urgent and obvious global tension is that between the different levels of government which are emerging. We are still living with the nation-states which, one to two centuries ago (in the West), took power from the city-states out of which they were formed. Even the new nations in the Third World have largely followed this model. The fact that this form of government has successfully held the main levers of power in many nations for several centuries has endowed it with the image that it is the natural form.

Indeed, our general groups – as a whole – tended to assume that the nation was the *only* natural form. On the other hand, when asked the question directly, two-thirds of individuals forecast the decline of nation-states (over the period to 2025).

Kenichi Ohmae, the renowned Japanese management guru, reflects informed opinion when he says (1995), 'the nation state – that artefact of the eighteenth and nineteenth centuries – has begun to crumble, battered by a pent-up storm of political resentment, ethnic prejudice, tribal hatreds, and religious animosity . . . nation states were created to meet the needs of a much earlier historical period, they do not have the will, or political incentive, the credibility, the tools, or the political base to play an effective role in the borderless economy of today . . . [they] are no longer meaningful units in which to think about economic activity. In a borderless world, they combine things at the wrong level of aggregation.'

In recent decades, as the various social and technological 'revolutions' have developed, the position of the nation-state – as the only legitimate power centre – has come under increasing challenge. In one direction, the smaller units of government, cities and increasingly communities, have demanded more power. They have claimed, with considerable justification, that this is necessary to deal specifically with local variations in demand for public services – of which there are an increasing number, as the nature of communities becomes more complex. This pressure for decentralization is likely to increase with the move to individualism with its clear emphasis on individual needs, which are most effectively dealt with at the local level. This has been compounded by the fact that voters, perhaps trying to implement their own form of power balance (or more obviously recognizing that political requirements are very different at national and local levels), have frequently chosen 'opposition' parties at the local level – almost inevitably promoting a clash between the two levels. On the other hand, many national governments have tended in practice to consolidate central power, often almost to the exclusion of local power.

At the same time, national governments are being increasingly challenged by the larger groupings they have chosen to join. These regional groupings of nations have typically been set up for economic advantage; as NAFTA recently has been and as, somewhat earlier, was the EEC. And they have tended to develop a momentum of their own, leading them to become regional governments in waiting, like the EU now. Additional stimulation for such agglomeration of nations comes from the increasingly global nature of the forces facing governments, not least those coming from the financial markets. From yet another direction, the single-issue groups, too, have become international. Groups like Greenpeace, Amnesty International and Oxfam all now raise their challenges on the international stage just as much as within the nation-state.

Not only are governments, in these ways, under siege from both directions but the newly emerging reality is that government in general (in all its various forms) now operates at a wide range of very different levels, from the local community to global supranational bodies. It is becoming impossible to pretend that all of these are, in some way, part of national government. Instead, each of these levels should be seen as having its own specific role in the spectrum of political power. A new natural order has yet to emerge fully. Nation-states are still tenaciously hanging on to almost all the powers they have gained over the centuries.

Power is now increasingly being distributed among different layers of government, from local to global. Tension arises as national governments, the traditional holders of power, attempt to halt the inevitable trends.

Nation-State Institutions

One further problem, which will increasingly result in tension, is that political institutions – at all levels – still tend to be dominated by the traditional supremacy of the nation-state, and this is reflected in the political processes across the spectrum.

At the community level, for example in city government, these processes are typically dominated by exactly the same political parties as at the national level and which unproductively repeat the battles which rage at the national level. Yet the roles of the two layers of government should demand very different approaches and there would seem to be very little synergy to be gained from running the two in tandem. The increasing recognition of community politics already recognizes this trend.

At the supranational level the problem is even more obvious. In this political arena, to assuage the fears of the national leaders, the normal democratic processes – enshrined in one man one vote – have been discretely sidelined. The European parliament is still largely powerless, although it is slowly garnering more and more power. The real power is reserved for the Council of Ministers, where policy is thrashed out between representatives of national governments (under the constant threat of their vetoes). Such deference to member states is undoubtedly necessary when the superstate is being created, but it seems highly likely that the resultant superstates will eventually flex their muscles against their members.

Political power is dispersing to different levels, from local to global. The main tension arises where national governments fight to retain their traditional powers against the almost inevitable encroachment of the other layers. It seems likely that new political frameworks will be needed to guide *co-operation*, rather than confrontation, across these emerging layers.

The United Nations

The United Nations is anything but united, and much less than the global authority it would wish to be, let alone anything like a global government. Thus, all representation in its assemblies and councils is strictly on the basis of the nation-states, exactly as its title implies. In theory the smallest nations have the same rights to representation as the largest. In reality, the General Assembly, where they are represented, remains largely powerless. Instead, real power is held by the Security Council which is by any standard grossly *un*representative, with a small number of nations determining the fate of all.

Even when the United Nations does decide to act, its actions may be blocked by individual nations, above all by the United States. After the demise of communism and confronted by the Iraqi invasion of Kuwait, it looked as if the UN was, at long last, about to fulfil the role demanded of it by the modern world. Unfortunately, as subsequent events demonstrated in Bosnia (where separate members of the UN Security Council held conflicting interests) and Somalia (where they had no worthwhile interest at all), this proved to be an illusion; it only occurred in the case of Iraq because there was a fortuitous coincidence of the interests of the United Nations with those of the United States; and the fragmenting Soviet Union had retired injured.

One area of conflict, therefore, may well be the replacement of, or updating of, the global institutions. New groupings of nations, especially in the Third World, will clash with the sitting tenants. It is likely that the United States will come into increasing confrontation with the rest of the world.

Community Government

The layer of government which should be most directly in contact with us is that of community government, often characterized as local or city government. In recent years such local government has too often been forced to play second fiddle to national government and too often been forced to dance to the latter's tune. With the increasing shift to individualism the importance of such local government will grow rapidly in a very different form of politics to that played on the national stage. The only prerequisite seems to be decoupling these political processes from national politics.

Global Politics

While the trend to multi-state groupings, and even to superstates, seems clear, the future of global politics is much less obvious. What does seem evident is the need for effective global regulation and intervention, to counter the global forces impinging upon government, especially in terms of macro-economics; and nearly two-thirds of individuals expect some form of supranational control to emerge by 2030, when – they agree – there will be a 'new world order'. The growing excesses arising from the powers of the overweening financial markets are forcing a recognition around the world that some form of action is needed, co-ordinated on a global scale. The very real threat, at present, is that any nation-state which dares to defy the financial operators will be destroyed by them.

Despite the very real advantages to be gained, politicians in existing nation-states are loath to hand power even to the multi-state groupings with which they share very obvious common interests and to which they have already committed themselves. It is difficult to see when such processes might be extended to nations with very different interests and which have often been the subject of centuries-long enmities! Bringing together Israel and its Arab neighbours has proved nearly impossible, even when they should share very clear common interests.

Global government, in the form of co-ordinated global actions, has already worked a number of times in the past. The allies in World War II put aside all their differences and worked well together, and with the politically very different USSR, across two quite different theatres of war. To a lesser extent, something similar happened in the Korean and Gulf Wars. As a peacetime example, the stability afforded by the Bretton Woods system has been credited with underpinning the golden age of growth between 1950 and 1973.

A separate problem is the presence of the existing global institutions. Unfortunately, these institutions – most notably the United Nations, but also the World Bank and IMF (also part of the Bretton Woods agreements, but more or less independent of UN policy since 1973) – are largely designed to bolster existing national differences. They do not meet the needs of most of the world and are becoming less and less effective. It seems likely that global politics will have to change significantly, to meet the changing needs of global government. The existing global institutions do not meet these needs – and will need, at least, to be drastically reformed.

Reactions to Change

The pressure for change will mount. Put in a more general context, the problem is that the existing global institutions offer a massive leverage to the most powerful nations. Despite the regular protests about their bureaucracies, this is why they are tolerated by the Western powers. As the power balance changes, however, this leverage will increasingly come to operate against those Western powers.

In the modern world (and especially in the post-modern one)

power follows the people and, in particular, the numbers of people. When, by 2025, more than 90 per cent of the world's population will belong to the nations currently categorized as being in the Third World, the small minority (10 per cent or less) in the current Western bloc will probably lose not just most of their influence but all of it. Changes will come when the Western powers recognize the inevitability of this process.

Unfortunately, at present the internal political processes of the Western powers put the priority on the short term and discourage strategies for the longer term. In addition, it is not clear how rapidly the United States will recognize, let alone come to terms with, the loss of its global empire. This may be the deciding factor in the West's response. When the changes will come about is unpredictable. Whether they will come about in time for the Western powers to sue for more favourable terms, before punitive ones are gleefully imposed by their erstwhile colonies, is also questionable.

Paradoxically, whatever the political outcomes, the West will already have given one important gift to the new world: its language. As Joseph Coates (1994) forecasts, with a considerable degree of probability, 'English will remain the global language.' This fact should not be dismissed as peripheral since language contains a great deal of hidden culture. In this way, whatever else happens, the West will determine much of the cultural inheritance of the future.

10

DARK FEARS

Previous chapters have described what have generally been optimistic views of the future. In this chapter we move to a very different aspect, one which is potentially much more pessimistic. It encapsulates the dark forces which have generated considerable pain and anguish in previous transitions. At the heart of these forces lie power structures dominated inevitably by the established political parties.

Although the 'optimistic' forces are most likely to dominate society in the longer term, the dark forces may make the transition painful. Previous transitions have been bloody as the establishment fought to preserve its position and privileges against the encroachment of the new, changed society. This time, however, bloodshed may not be inevitable. It was not just the demise of Communism and the dissolution of the Soviet empire which were accomplished peacefully – very much against the previous precedents for such revolutionary changes – but there have been equally far-reaching but peaceful changes which have occurred in South Africa, Chile and Argentina, where right-wing regimes have handed over power to the masses. Just a decade or so ago, when global nuclear annihilation was still a very real possibility, the arguments of those pundits who predicted the end of civilization would have been much harder to counter. Now there is considerable hope that revolutionary pains may disappear and transitions take place without dramatic bloodshed.

A number of other authors have recognized the revolutionary nature of the times in which we live, typically using the previous Industrial Revolution as their model. Alvin Toffler, for instance, has given his books titles such as *The Third Wave* and *Future Shock* to emphasize the all-encompassing nature of such changes. Technological change is indeed endemic, and will have massive impacts on some aspects of the future of society. On the other hand, we have now reached the stage where the resources available to humanity can meet almost all its reasonable needs, and in effect are now unlimited. In this context, such technological potentials have become merely the supporting factors for the new revolutions which are in the process of confronting a largely unsuspecting world. These new waves of revolutionary change, described in the preceding chapters, are to be detected across all aspects of society and especially in its social organization.

While the forces we have examined, from technological changes through to the remaking of society as a whole, may be fundamental in the longer term, those operating in the political environment are more immediate in terms of their impact. They determine how, and when, the underlying changes will be allowed to occur. They determine where and when the stresses building up are to be released, and the longer they are contained the more powerful the ultimate shock will be. These political forces are often hidden from view, especially in Western societies. As a result, the forces themselves are poorly understood, not least by the very politicians who would claim to control them, and the changes which are occurring around them are very much misunderstood. The mismatch between the changed circumstances and the unchanged policies for dealing with them is what often causes the worst revolutionary pain.

Governments, in their anxiety to deal with new forces they do not understand, often grasp at 'solutions' which make the problems worse. For example, choosing a strategy based solely on maintaining a low rate of inflation during a time of great change is at best naive, even if all the other simplistic nostrums (from control of money supply to exchange rates) have failed. The precedents would suggest that, all other things being equal, inflation should normally rise to reflect the uncertainty about the future. The deliberate employment of draconian measures such as unemployment to reverse this natural process in order to achieve unnaturally low rates of inflation, at a time when the disadvantaged are already suffering from the problems of transition, is inhuman and ultimately counterproductive.

The Structures of Power

Due to the cultural context within which they operate, these power structures are, unfortunately, largely invisible to Western commentators and, in particular, to Western managers at all levels. They tend to be incorporated as basic assumptions, hidden underneath management theory as much as that of market politics. At one extreme the market is supposed to mediate impersonally (and most effectively) between individuals. At the other, democracy allows those individuals political control over the few remaining aspects of life which the market cannot address. In recent years it is these two pillars – market and democracy – which have been chosen to symbolize most of what is important to Western political systems, rather than the capitalism which had previously been said to distinguish these systems.

Perhaps the greatest change has been in the demands placed upon the leadership role of government. Where, previously, it was required to give political (almost spiritual) leadership, with a clear vision of a nation's chosen future, it is now expected to undertake capable management of the nation's resources. Now we find most of the solutions to our problems ourselves; and governments are fooling themselves when they think they are still running a hierarchy which is in control of such changes.

Since the traditional structures are disintegrating and can no longer offer such comfort, the latest 'industrial revolution' poses very different challenges. Traditional structures – overcome by the forces of change – are being torn apart. The symptoms are already to be seen around us, not least in the marked instability in voting patterns.

The electorate, across much of the world, is beginning to see that their political masters, no matter how much they bluster, are lacking workable policies to meet the new challenges. As Fintan O'Toole (1995) suggests, 'What characterizes our approach to the Millennium is that the ideologies of both left and right are in trouble . . . the rise of Green thinking has drawn attention to the physical limits on progress and to the very real prospect that [it] could lead to oblivion.' *The Economist* (27 November 1993) also highlights the problems faced by the opposing ideologies, though it ascribes them to more complex causes, pointing out that 'voters in established democracies are showing signs of an increasing readiness to desert the familiar parties of old . . . for years, political parties in old democracies have been struggling. . . . In America, as in other countries, society has been growing less cohesive, more granulated. Education, switching jobs and moving up the class ladder have made people less inclined to take their politics from their parents and more inclined to form, and change, their own political opinions.' Whatever the reason, the structures are being torn apart.

Despite the euphoria of the victory over their enemies, most notably characterized by Francis Fukuyama (1992) as 'The End of History', Western leaders now need to learn the more basic lesson and view their *own* political future with just as much trepidation. The same structural forces, which should perhaps be better encapsulated as 'The End of Ideology', are now undermining the position of Western governments, as previously they did those of their Communist counterparts. In addition, an end to the ideological battle between communism and pluralism, each of which was long presented to supporters in stark black-and-white terms, will make the specific principles of 'representative democracy' – which is now the only real political offering in much of the West – look increasingly unsatisfactory. The supporters of this now exposed ideology will in this way be forced to defend their choice with rational arguments rather than the quasi-jingoistic pleas on which they were previously able to rely. *The Economist* adds that: 'In the absence of its competitor [communism], the consequence is a weakening of ideology in general.'

Driven by the changes wrought by the latest industrial revolution, a parallel political revolution is sweeping the globe – communism was just its first victim. Although it is not yet clear what exact form they will take, almost all nations will experience massive shifts in political power. Crises of legitimacy will increasingly feature on the political scene.

Democracy

Let us, therefore, look at the key assumptions supposedly underpinning much of Western society; and let us start with its proudest boast, that it alone knows the secret of true democracy. As early as 1780, Edmund Burke, in a speech on economic reform, made the definitive statement that 'The people are the masters.' It is, thus, assumed that the Western 'democracies' have long-since found the means for 'government of all the people by all the people for all the people', and the two concepts, of democracy and freedom, have long been so intertwined in the West as to be almost inseparable.

Surprisingly, in view of the hype of past centuries, the United States' Constitution actually has surprisingly little to say as to the individual's rights in this context; and even the rather more idealistic Declaration of Independence has nothing to say beyond the one, most memorable and so often quoted, sentence seeking the rights of 'Life, Liberty and the Pursuit of Happiness'. Even so, this model, of government separated from the people, but held in check (only of the grossest excesses), still remains a basic assumption of Western 'democracy'. Thus, Andrew Adonis, correspondent at the *Financial Times*, and Geoff Mulgan, director of Demos, are able to state (1994) that 'modern government is exclusive and élitist. It generates unreal and largely ignorant expectations on the part of voters, and encourages political élites to trade simplistic, cut-and-dried solutions to problems as the currency of electoral politics. Political alienation and ignorance are systemic.'

The patterns of power are about to change, however. Alvin Toffler (1980) makes the fundamental point about the new political realities, 'The first, heretical principle of Third Wave government is that of minority power. It holds that majority rule, the key legitimating principle of the Second Wave era, is increasingly obsolete.... In place of a highly stratified society in which a few major blocs ally themselves to form a majority, we have a configurative society – one in which thousands of minorities, many of them temporary, swirl and form highly novel, transient patterns.' He adds his own solution to the problem which 'lies in imaginative new arrangements for accommodating and legitimizing diversity – new institutions that are sensitive to the rapidly multiplying minorities ... "semi-direct democracy" – a shift from depending on representatives to representing ourselves ... if our elected brokers can't make deals for us, we shall have to do it for ourselves.' Therein lies the potential confrontation, between the old and the new, and possibly the greatest danger of a painful transition to come.

Most interesting of all, in terms of my own experience with attempts at defining democracy in the context of Ethiopia, was the fact that most of the Western governments involved apparently failed to recognize that what they were promoting was just one possible approach to democracy. In particular, they were unwilling to recognize that in practice it might be less than perfect.

Such absolute government by an effectively closed élite might have been justifiable, in practice if not in theory, when most of the population were ill-educated and even more poorly informed and might not have been expected to cope with the complex issues to hand. A hundred years ago fewer than 2 per cent of Americans went to university; now more than a quarter do. In addition, through the almost universally available medium of television, people worldwide are remarkably well informed. The ground has not merely moved, it has been destroyed by a political earthquake of monumental proportions.

The leaders of the multinationals we have talked to believe that government is now about effective management of day-to-day events rather than political visions on the grand scale; it is 'learning to apply small fixes for small problems'.

Despite the many critics of his pragmatic 'flexibility', which many see as a cynical approach to retaining office at any cost, it is even possible that Bill Clinton might be the model for future leaders. He adjusts his position to match the demands of the electorate more or less exactly as truly representative democracy would in theory require. As a result, he has also distanced himself from the traditional party machine as post-modern politics might demand. Just why he has adopted this approach is not obvious. His critics may be correct in claiming that it is a cynical manipulation of the electoral processes, though it is not clear why giving the electorate what it wants is undemocratic. Much the same could be said of Tony Blair in the United Kingdom – though he has, at least, taken most of his party with him (albeit kicking and screaming almost every inch of the way).

For the record, therefore, Western democracy, and most forms of government around the world, at best only offer a limited check on what government does and even then only by seeming to offer the opportunity to reject existing political power at intervals of five years or so. It is that very opportunity, to reject existing political parties, which is now being taken, with some enthusiasm, by not a few electorates.

The democracy so proudly claimed by Western governments is just one quite limited example of a wide range of options available. In terms of any legitimate claim to represent all of the people, it is now a rather ineffective form. It will become increasingly difficult to persuade electorates that their only involvement in government is, at relatively long intervals, in terms of being able to enact checks on its grossest excesses through the ballot box.

Competition

Less obviously, perhaps, the Western concept of democracy is based not upon the co-operative ideal enshrined in 'all of the people' but upon the contrary notion that politics, along with many forms of public life, demand a contest between opposing ideas, of which only one can be the winner. This concept was previously seen in its purest form in the confrontation with communism; then Western democracy was most easily defined in terms of its opposition to the dark forces of that communism. It is still enshrined in 'party politics'. In the traditional ideal, this has just two equally matched contestants (Democrats and Republicans or Conservatives and Socialists), whose opposing views are placed before the people in a perverse form of beauty contest. The major Western powers simply cannot cope with the concept of co-operative government. It may come as something of a surprise to their electorates, but in Western political eyes democracy is defined simply in terms of the existence of competing political parties. The wider and more bitter the divisions between them, even if these are artificially created, the more democratic is the process considered to be! Even more surprising is the fact that the concept of 'government by all the people' seems to play no significant part.

Co-operation

Some years ago, some colleagues and I made the fundamental observation that – in society – 'forces *add* in conjunction and *subtract* in opposition'. We believed that this should be self-evident, and still so believe; nothing has happened since to change our minds – indeed the obvious failures of government policy in the 1990s have reinforced our argument. Despite their failures, however, most politicians still claim exactly the opposite!

With the end of the global ideological conflict the one remaining public confrontation has had to be engineered by those who seek confrontation. It is that between those, usually of the right, who believe that the most important thing is to make the economy work as efficiently as possible and those, usually of the left, who prefer to concentrate on looking after the unfortunates who get the least from the economy. This confrontation between left and right is becoming much less relevant to most of the electorate. Competition for scarce resources in this way makes little sense when such resources are no longer scarce. It makes much more sense to co-operate with others to create the better society that you, and they, desire; it should be possible to create wealth efficiently and share it.

What has been forgotten is that the 'market' is basically a device for allowing individuals to co-operate. It allows each of us to focus on what we do best, so that society overall works most effectively. According to Michael Lerner (1996), 'Most Americans hunger for meaning and purpose in life. Yet we are caught within a web of cynicism that makes us question whether there could be any higher purpose beside material self-interest and looking out for number one. . . . People treat one another as objects to be manipulated.'

Fortunately, the linkage between the individual and society is now widely recognized, among the electorate at least, if not among all politicians. It is only the most dogmatic politician, such as Margaret Thatcher, who would consider saying that 'there is no such thing as society'. In addition, our own research shows that even among those supposedly most committed to competition, private-sector business organizations, co-operation rather than competition is now the rule in the wider community. When confronted directly by major competitors, half of all organizations choose the route of co-operation and, in the acid test, only a third said that they price their products with competition in mind.

In the face of such widespread common sense, it represents a major triumph of some sort for Western politicians to have persuaded so much of the world that the only acceptable political and economic process is (still) that of competition.

Participation?

Having earlier put on record some caveats about the otherwise relatively peaceful move of power from the élites to the individual, it seems reasonable to say that at long last the masses (operating as informed individuals using the ballot box rather than an ignorant proletariat storming the barricade) are coming to demand their rights in government and, in particular, their rights to participate in it.

Participation – often now described as direct democracy as opposed to (traditional) representative democracy, and described in more detail later in this chapter – may be a potentially fraught solution to put into practice rather than to describe in theory. After all, it is no longer possible to have state decisions directly decided by everyone, as it was, to some degree, long ago in the golden age of Athens – though some might recommend an electronic (Internet) equivalent. Instead, there may be a spectrum of opportunities for involvement. At the upper end this consists of the whole panoply of existing political appointments, from members of the inner circle of government down to the lowliest town councillor or school governor and those enjoying the patronage of government as members of non-elected bodies. At the opposite extreme is simply a belief that government is being undertaken on behalf of everyone, though in reality it often governs in the interest of special groups.

Clearly, the supporters of the winning party expect to gain something from their support, but there are many other groups which are almost as successful in diverting power to their benefit. The Mafia in Italy (and to an extent in the United States) pull many strings. Surprisingly, considering its small numbers (of just a few per cent of the population) and its constant cries of discrimination against it, the Jewish community in the United States has had a remarkable degree of influence on government there – vastly out of proportion to its numbers and achieved by an equally remarkable degree of unity of purpose (especially in targeting congressmen and senators with threats, bordering on blackmail, of electoral destruction).

It is usually clear that government is biased in favour of a chosen few and against the majority. Bribery of key supporters might have worked in times gone by, but now it merely heightens the sense of moral outrage experienced by the disadvantaged majority.

The basic requirement for participation is that the population as a whole can comfortably accept the assertion that government really is working on its behalf. Beyond this, fuller participation will come when the political establishment opens its ranks to anyone with the capability, and the desire, to participate in the processes of government. Such a change is easy to make in theory, but almost impossible to make in practice when all the gatekeepers have a vested interest in maintaining the power of the élite. The forces for such change will come from the demands for legitimate representation of the views of individuals. When failure to respond to such demands is seen to undermine the legitimacy of government, it will act – almost certainly too little and too late.

Probably the most distinctive feature of the Western model of democracy is its emphasis on competition between powerful (opposing) political parties. Where, as with the demise of communism, the only significant competitor has been disqualified, such a system is immensely weakened.

New models of democracy, based on co-operation, are needed. They are currently being explored by a number of countries, including some of those emerging from the shadow of Communism (which promised such co-operation, but rarely delivered it), but no widely accepted or demonstrably viable model has yet emerged. The key change, which most governments will probably concede too late to avoid considerable tensions, will be an obvious desire upon the part of government positively to involve the electorate.

Pluralism and Consensus

In the post-war years, the practical workings of Western democracy were often based upon processes lying outside of the formal electoral system. Thus, in the 1960s and 1970s it came to be generally accepted that there were many groups which also wielded a degree of political power. The belief then was that it was possible to achieve a balance among these groups, or at least to reach some form of consensus. In the 1980s the political scene had returned to confrontation; consensus was no longer sought and even a search for consensus was seen to be a sign of weakness. Electorates, finally disillusioned by the failures to deliver the certainties they were promised in the 1980s, now seem to be moving back to a pluralist consensus.

Pressure Groups

In the earlier post-war decades an almost essential aspect of such pluralism was the existence of pressure-group politics. Each of these groups attempted, outside the normal democratic processes, to sway the decisions of government in its favour. It is still possible for outsider groups to have a significant impact, if their pressure is well enough managed. In the United Kingdom, the anti-roads lobby, which has been especially successful, brought together a whole range of groups – from the 'Dong Tribe' of travellers who demonstrated violently, to Transport 2000 which published well-researched briefs – all of whom worked together for the common cause, possibly setting an important precedent for lobbyists and activists in other fields. Elsewhere, the Catholic Church has used the Pro Life movement to apply political pressure in support of its views. The point is not that such pressures are good or bad; it is simply to recognize that they exist. Pressure groups can wield significant political influence, although their impact may be variable and unpredictable.

Single-Issue Politics

Pluralism has seemed to re-emerge in an even stronger guise: as a range of single issues on which groups within the population have focused to great effect. The commitment to such focused politics can be more intense than that given to party politics since it is not diluted by all the other issues about which the individuals within them feel less strongly. The commitment is often much more intense in terms of the numbers of individuals actively supporting such movements and in terms of those especially highly motivated activists running the campaigns. It is much easier to give a high level of emotional commitment to a woman's right to an abortion, or to an animal's right to live without pain, than it is to have the same visceral reactions to the compendium manifestos adopted by the traditional, national political parties. Peter Drucker (1993) states the position in stark terms: 'Parties are in tatters everywhere. The ideologies that enabled European parties to bring together disparate factions into one organization to gain control of power have lost most of their integrating power.... Governments have thus become powerless against the onslaught of special-interest groups, have indeed become powerless to govern.'

One result of the disintegration of traditional social groupings and, accordingly, of traditional party politics is that single-issue politics is now growing rapidly. Indeed, it would seem that many of the individual voters are now increasingly committing to a portfolio of single issues which, in aggregate, define their overall political position.

What have yet to emerge are genuine portfolio parties. These would seek to offer attractive combinations of policies for such single-issue voters, as Bill Clinton did in 1996 (working in opposition to his party) and Tony Blair did in 1997. By this means, such new parties may hope to increase their electoral coverage sufficiently to achieve a majority. In the context of portfolio parties, the sophisticated marketing research techniques already available – essentially using factor and cluster analysis – could tell the builders of such parties whom to appeal to and how. This does not, however, answer one major potential problem – that the content of the most attractive portfolio might change significantly over time, posing problems for any party trying to establish a unique identity.

With the decline of traditional voting groups, single-issue politics are likely to grow in importance. Individuals may increasingly come to build their voting behaviour on portfolios of such single issues.

Possible Political Solutions

The electorate is increasingly aware that significant changes are afoot in the political processes, at all levels and worldwide, although, as yet, they do not appreciate exactly what these changes might be nor the true scale of them. The one group, blinded by a global version of groupthink, which is determinedly unaware of them is made up of the various sectors of the establishment, and in particular the politicians, who are charged with implementing such change.

A prerequisite for any evolutionary solution to the emerging problems is that those politicians are persuaded to recognize the changes which are already taking place around them. Unfortunately, history tends to suggest that politicians will only recognize the facts of life when it is their political life which is imminently at risk, when revolution threatens. Recent history suggests that this recognition need no longer wait for bloody revolution, but when all the various changes will come about is much less predictable. It seems likely that those changes which will apply at the local level will arrive earlier; those at the global level will take longer, even if they are more obvious.

The most immediate change is simply the recognition that a new model is needed, based upon co-operation rather than on confrontation. The essence of this can be encapsulated in the concept of zero-plus (where both parties in any negotiation can reasonably expect to gain from co-operation) as against that of zero-sum (where one can only gain at the expense of the other) which currently rules. Regrettably, zero-sum confrontation is a model which Western politicians seem unwilling to relinquish.

A second prerequisite is that the information-carrying capabilities of these political processes should be dramatically improved. In the days, long ago, when the current frameworks originated – graphically still encapsulated in the rule that the opposing parties in the British House of Commons should be kept two swords-lengths apart – the electorate were literally isolated in their ignorance and the politicians were part of the élite which alone was privileged to know what was happening in the wider world. Now, paradoxically, the reverse is true. We, informed by round-the-clock and round-the-world, television (often across a multitude of channels), cannot avoid exposure to the wider world. Politicians, on the other hand, increasingly live in an artificial world which often bears little resemblance to the real world of their constituents. Isolated by the barriers they have thrown up, they are the prey of officials and lobbyists and of their fellow politicians, who comfortingly share their own views.

Referenda, Juries and Opinion Polls

There have been many solutions offered to bridge the communications gap between politicians and their electorates and to offer the latter more influence. *The Economist* (17 June 1995) asks the rather provocative question: 'When the public view can be tested so frequently and easily . . . why have elected representatives at all?' and goes on to pose an even more fundamental one: 'Will representative democracy prove to be merely a 200-year intermediate technology, a bridge between the direct voting in ancient Greece and the electronic voting of modern California?'

Experience has already shown that referenda can be used for major, crucial issues. They work, even if politicians do not like to relinquish power in such a manner, as long as the electorate is well informed and not *mis*informed. The evidence is that, in general, a well-informed electorate will approach such referenda with great responsibility. It seems likely, therefore, that their use will grow, especially as the cost of running them will decrease with the growth in the use of computer technology in this area.

But referenda are not the only possibilities. There are a number of other devices which could also be used to encourage direct democracy, and two-thirds of our individuals expect some form of direct democracy in general to be prevalent by 2030. One of these, which has a long record of success, is that of the jury. If we have been willing to entrust the life of the defendant to a small group of his or her peers – and the evidence is that this works well – it should not be impossible to use such processes to deal with lesser issues of public policy. Apart from Tony Blair's 'advisory' panel of several thousand voters, there are no obvious moves afoot in this direction in the major democracies.

Beyond these, the more public use of marketing research could, and eventually should, improve the communications process. It is true that opinion polls have recently earned a rather dubious reputation, but properly handled (and once they are respected by those being interviewed), they can give accurate insights into what the electorate in general is thinking. Where you are building a consensus, such knowledge is invaluable – even if it is only used to map out the changes in opinion you wish to make. Many of these devices are scarcely revolutionary in nature, and could happily operate within existing frameworks. Thus, some of the various forms of proportional representation would allow the voices of almost all the electorate to be heard, including members of minority groupings rather than just the favoured majorities. It seems likely that such approaches will spread from the nations which already use them, with great success, to the laggards (which, unfortunately, include much of the US and the UK).

Despite the emotional objections of the politicians who do not wish to dilute their power, all of these changes are about communication rather than political ideology. Even politicians may seem to agree with this; for they often claim that it is the 'presentation' of their policies which is at fault. What they fail to understand, however, is that now the most important communication is *from*, rather than to, the electorate. Above all, they must stop talking, and *listen*. Allowing small groups and, increasingly, individuals to communicate with government is likely to become central to the whole political process. With the advent of widespread computer conferencing, it has now become possible for even the most remote community to maintain regular contact with its representatives.

A political revolution is under way. The fall of communism was just one step towards the emergence of new political frameworks and processes across the globe. The traditional political parties are far too isolated from their electorates, which are increasingly demanding their rights by other means – not least through single-issue groups. These may, in turn, lead to the emergence of portfolio politics, which will demand a listening government. The prime requirement is a change of attitude, so that the government *wants* to listen.

11

REVOLUTIONARY PAINS

The confusion and uncertainty caused by so many revolutions coming along at the same time, from the 'IT Revolution' through 'Post-modernism' to the 'End of Ideology', have resulted in a great deal of pain for some of us. The first Industrial Revolution was also accompanied by political revolutions around the world. In Thomas Kuhn's famous words from the field of science, a paradigm shift is under way. According to his observations, the result is almost always a period of great uncertainty as the defenders of the old world order, the old paradigm, dispute with those of the new. We are currently in the middle of such a paradigm shift, probably the greatest paradigm shift of all time. The pain which accompanies this is, therefore, understandable.

More than a quarter of a century ago, Alvin Toffler predicted the problems we would be facing in his book *Future Shock*. In that influential work he said, 'I coined the term "future shock" to describe the shattering stress and disorientation that we induce in individuals by subjecting them to too much change in too short a time.' He went on to make the related point that, 'In the most rapidly changing environment to which man has ever been exposed, we remain pitifully ignorant of how the human animal copes.' A quarter of a century on, Toffler's predictions have come true, but we still do not know how we will cope. More surprisingly in view of the impact of his original book, most of us still do not even recognize the symptoms he predicted for what they really are. Indeed, our leaders often are unwilling to admit that such stresses exist.

These themes have also been central to much of Toffler's later work. For instance in *Powershift* (1990) he added, 'The simple speed-up of events and reaction times produces its own effects, whether the changes are perceived as good or bad. It is also held that individuals, organizations and even nations can be overloaded with too much change too soon, leading to disorientation and breakdown in their capacity to make intelligent adaptive decisions.' I can't explain the phenomenon of 'revolutionary pains' in better terms than his, yet despite the proven accuracy of his previous predictions, most of our leaders still remain obstinately blind to what is happening around them.

We noted at the beginning of the book the confusion which, as a result of these pains, pervades the 1990s. The seminal event of the decade actually took place at the end of the previous decade, in 1989. The fall of the Berlin Wall signalled, for many of us, the start of a global revolution in politics. Reporting the research results of the internationally respected Henley Centre for Forecasting, Stewart Lansley (1994) stated that 'The crisis of the nineties is not confined to the problem of political and economic adjustment in Eastern Europe nor to the impoverishment in large parts of the developing world. Despite claims of the final triumph of western liberal capitalism, the democratic west is facing its own deep-seated crises of identity and management. Western values seem increasingly out of tune with the needs of modern society.'

As expected, the establishment is desperately fighting against the tide of history. For the 'crisis decades', using Eric Hobsbawm's phrase, of the late twentieth century, they have been adopting ever more reactionary measures to try and return to earlier and 'better' times. They had some success during the 1980s in particular: governments around the world succeeded in taking many of their voters with them, 'a silent majority returning back to basics'.

Many politicians around the world have chosen to focus on the one simple economic measure they believe they do understand – inflation. As we saw in the last chapter, the test of good government has now become its ability to keep its economy in a mild form of recession.

As we realize what is happening, we are already starting to move ahead of our governments. It has to be recognized that the politicians still have their supporters in other parts of the establishment. In particular, despite the new realities of employee power and of the massive investment in people, too many senior managers unwisely used the recession at the beginning of the 1990s to reduce their workforces ruthlessly. The 3Rs – Restructuring, Re-engineering and Redundancy – came to dominate Western management thinking and took their toll on investment in people. At precisely the same time, the comparable Japanese 3Rs were Restructuring, Re-engineering and Retraining! They, at least, never lost sight of the importance of their workers. Our research indicates that it is the Japanese corporations, rather than the Western governments, which will be proved correct in the longer term.

The Isolated Establishment

Alvin Toffler (1980) explains that 'Second Wave elites fight to retain or reinstate an unsustainable past because they gained wealth and power from applying Second Wave principles, and the shift to the new way of life challenges that wealth and power.' It now seems to be an inevitable feature of almost all governments, communist just as much as democratic, that they are composed of a self-perpetuating élite. This aloof élite extends beyond politics to encompass all aspects of power: business, the law, and – too often – even trade unions. It is the problems of the establishment, rather than those of the masses, which have led to many of the dilemmas being encountered in the crisis decades at the end of the twentieth century.

In the long term the establishment will be unable to resist change. In the shorter term its response will determine whether the transition is painful or even as bloody as previous ones. I believe, or at least hope, that the new battles will be fought with the ballot box rather than the gun.

As a final optimistic footnote to this section, Francis Fukayama, in his sometimes misunderstood treatise *The End of History* (1992), notes that 'there was a remarkable consistency in the democratic transitions in Southern Europe, Latin America and South Africa. Apart from Somoza in Nicaragua, there was not a single instance in which the old regime was forced from power through violent upheaval or revolution. What permitted regime change was the voluntary decision on the part of at least certain members of the old regime to give up power in favour of a democratically elected government.' The even more dramatic revolutions in Eastern Europe were just as bloodless. Perhaps the modern establishment is different and we will not see blood on the streets this time.

Rejection of Politics

The outcome of the revolutionary changes in the political views of the electorate has not yet been that the masses have risen up against the establishment in bloody rebellion. Instead, in the West the population has, while learning to use their new powers, simply rejected politics. Possibly as result of the élite's tight control over the processes of legitimization, the public at large has not seen revolutionary change as a viable option. Unlike earlier revolutionaries, they have not been desperately hungry. Instead, they have been rather comfortable. The result is that they have, with some notable exceptions, chosen to distance themselves from the whole political process while gradually developing a quite considerable distaste for all of those involved in it (journalists almost as much as politicians).

What has been developing is a growing awareness by individuals of their new powers in the political processes. As one example of this, referred to earlier in terms of single-issue politics, many contentious issues have merely been taken outside the normal frameworks to be dealt with in specialist actions by specialist groups. More generally, however, electors are beginning to use – with increasing effectiveness – the power of the ballot box.

Sometime in the 1980s it did appear as if a large proportion of the electorate started to vote tactically. By 1997, in the United Kingdom this was perhaps evidenced by the electorate's virtual destruction of the Conservative Party. But there have been similar outcomes around the world. In France, Italy, Canada, Japan – to name but a few – there were massive switches in voting patterns which heavily penalized the sitting tenants.

The result can be seen as a form of 'anarchy', using the term in the strictly theoretical sense of 'an absence of government' (which I will use throughout the rest of this chapter) rather than its more popular use as 'political disorder' in general. It is, indeed, an especially pure form of anarchy since the electorate is only concerned with destroying the power of government. Such anarchy might now be made workable, in the sense that although the traditionally rigid hierarchies of government are already disappearing, there is usually still a broad consensus among the majority of the population on the overall package of policies to be adopted. The danger potentially comes from those outside this mainstream consensus. There have always been groups which have been disadvantaged, as against the majority groupings. The history of the United States, as just one obvious example, has often consisted of the struggles of such groups to join the mainstream: the Irish, the Italians, the Blacks, the Hispanics. Around the world the struggles of such groups have often led to terrorism – the IRA in Northern Ireland, the Palestinians in Israel – and may continue to do so.

To these can now be added the impacts of the new differences between the 'underclasses' and the mainstream. In the United States these latest divisions have resulted in a significant part of the population being, in effect, disenfranchized from the consensus which is the driving force behind the rest of society and, worse still, seeing no possibility of redress (since the majority, almost as much as the politicians, has turned its back on them). The terrorism generated as a result is often not obvious, but is just as destructive none the less. With nothing to lose, members of the underclass are opting out of the values which power the consensus, to build their own state within a state. Not least, crime has become a staple industry. It has taken a generation or more to create this situation in the United States, and it will take at least as long to reverse it even with the most committed policies. The outcome of similar Sicilian resistance in the form of the Mafia (which originated as a self-protection group) still haunts us.

Around the globe, as electorates recognize the failings of existing systems, the first reaction is to reject government and refuse to participate in its legitimization. In a subsequent step, electorates are choosing to punish and even destroy governments by visiting massive electoral defeats upon them. When even this punishment fails to work, it seems likely that more direct action, probably in the form of new quasi-political groupings, will emerge. These will probably be committed, however, to put in place new democratic processes. The essence of the political future over the short term may be a new but workable form of anarchy. However, the delicate processes involved may be derailed by the confusion arising from the conflict with the underclasses.

'Dark Fears'

Despite the fact that this is the only part of the book which clearly has the potential for pessimistic outcomes, in practice, even here, the balance probably swings towards optimism. The dark forces are those which might delay or possibly destroy the chances of a happy future for humanity by opposing, or at least threatening, short-term implementation of some of the positive developments described in the earlier sections. Certainly, they will determine how serious are the revolutionary pains which accompany these changes. They are almost entirely political in nature.

The biggest threat, as I have already suggested, is that the revolutionary pains being experienced, especially by the establishment itself, will not be seen as symptoms of underlying changes but will themselves be seen as the problems and the targets for government action. This happened, to a degree, in the 1980s and 1990s and led to reactions which greatly magnified the pain – especially in terms of unnecessary unemployment. In previous centuries, such revolutionary pains escalated to cause widespread bloodshed and this must still be a fear hanging over the current problems. Yet, there is evidence that because modern populations are so much better educated and informed, they will be able to handle the transitions much more effectively and, equally, that members of the modern establishment are more willing to cede power more peacefully.

12

SHAPING THE FUTURE

You may consider that many of the ideas which have emerged from our researches are self-evident. On the other hand, you may also consider that some are highly speculative; it is up to you to judge their validity in the unique circumstances which you will encounter. In this spirit, therefore, over the next few sections I will summarize what I consider to be the most important themes which have emerged from the scenarios. In the latter part, I will briefly suggest how the future, in general, might be managed: by individuals such as you and me; by organizations such as those we work for; and, in particular, by governments which we vote into power.

Positive Optimism

The key to the future will be our own attitude. Unfortunately, even as the Millennium approaches, there still seems to be some public pessimism around; though our most recent results indicate that this may now be largely confined to the media and the media fuels this gloom by an 'amplification spiral' of increasingly alarmist reports. On the other hand, our own research results have revealed, over several years, that there is significant private optimism, albeit that this is too often well hidden from the media. Thus, even our earlier (1993) directly quantifiable results recorded a significant degree of optimism in line with our own subjective interpretations of the discussions which took place in most of the research groups; and our latest (1997) results are even more positive. According to expectations theory, these optimistic expectations of our groups – unaffected by the prevailing moral panic reported by the media – should indicate an equally optimistic and generally happy future for all of us, across the globe.

Unlimited Resources

The most important assumption behind this book is that our future no longer has to be constrained by resource limitations. Our future will be shaped by the choices we make, or at least by the decisions governments take on our behalf.

One of the central developments in the provision of unlimited resources will derive from the medical sciences, in the form of significantly increased longevity. We may soon live for a century or more, on average; and this will result in our retirement age being increased to 70 years and beyond, adding to our healthy working lives and adding significantly to the resources available to society.

Most of the technological changes will take place without government intervention. On the other hand, the greatest long-term development of physical resources will come from the colonization of space; and this will almost certainly need to take place on a scale which only governments will be able to justify, and even then it will almost certainly require international co-operation.

In the shorter term, the main driver for technological change will be the developments in the field of IT and in computer communications. These will revolutionize how we work and play, and how we relate to others. They may even lead to a further stage of evolution.

Individualism

Many of the greatest challenges will come in changes which affect how society itself operates. Of these, the most fundamental will come from the breakdown of groups resulting from the empowerment of the individual, newly released to fulfil his or her total potential.

One of the key developments is the emergence of women as a power in their own right and the shift across almost all parts of society to more feminine values, especially in terms of co-operation rather than competition.

Perhaps the greatest unknown, one which clearly worries governments in the West, is the breakdown of the nuclear family as we now know it. New forms of the extended family are already emerging to replace it, although society as a whole has yet to recognize these as desirable new developments. This problem is also at the heart of the search for the new community. Developments in new lifestyles will represent one of the most obvious external signs of social change; we will increasingly come to demand a portfolio of such lifestyles to match the specific circumstances in which we find ourselves.

Globalism

The emergence of a truly global 'village' will have major impacts because most of the individuals who will then hold the economic and political power will emerge from the nations currently in the Third World. One of the crucial developments will be the weakening of the nation-state, to be replaced by a spectrum of political structures at various levels upwards from the community to supranational groupings. Many issues, especially those relating to financial markets, along with communications and environmental protection, will demand management on the global scale.

The emerging power structures will inevitably favour the largest nations, especially China and India. Among the existing leaders, the

EU will emerge as a major force, as will Japan (at the heart of the Pacific Rim equivalent to the EU). The United States, however, will probably take a different, potentially apocalyptic, line.

Dark Forces

Symptomatic of the scale of the changes being experienced is the breakdown of political structures around the world. Political parties everywhere face extinction, to be replaced by new forms of representation, including various kinds of direct democracy.

Revolutionary Pains

The final point to be made is that the current levels of uncertainty, which are causing such pain and anguish are *temporary*! It ought to be expected that during a period when a number of major 'revolutions' are approaching their peaks there will naturally be a period of revolutionary pain as the new frameworks are put into place.

The real danger recognized by only a few commentators, is that as a result of the confusion the establishment in general, and that of politics (and the media) in particular, is getting ever more out of step with the rest of the population. Much of the establishment is retreating ever further into a nostalgic exploration of past solutions which, almost by definition, will not meet the problems posed by the future. As the gap between the establishment and the rest of the population grows ever wider, the potential for conflict between the two may escalate to dangerous levels. The one redeeming feature, however, is that eventually the establishment position will collapse, much as it did in Eastern Europe at the end of the 1980s. That collapse itself may lead to chaos, riding on a wave of popular euphoria, but that too will only be temporary, again as it was in Eastern Europe.

Managing our Responses, Managing the Future

I shall conclude this book by looking, again briefly, at how we may respond to the challenges posed by all these revolutions. Using a combination of the various theories we have developed and commonsense, I will work upwards from what we as individuals might do, through how our organizations may shape their own futures, to what governments might do.

Individual Empowerment

The first thing we have to recognize is just how powerful we have become as individuals. But with this power comes a degree of responsibility; for now we, individually, hold the future in our hands. Perhaps the most important requirement is simply the public acceptance that our view of the future should be optimistic. It then demands that we persuade others, and in particular our governments, to accept the same optimistic viewpoint. The alternative, a pessimistic decline into a new dark age, is, just as simply, unthinkable.

When it comes to individualism, the long-term challenge for each of us is to face up to the uncertainties implicit in what you have read in this book and to create the new role(s) demanded of us. The shorter-term challenge will be to fulfil our potential, and that will almost certainly demand a commitment to ongoing education.

The future of the family is one issue which can only be resolved by us as individual members of that family. You cannot legislate to create happy families, although you can persuade society to be more tolerant of the new forms of extended family which are even now emerging. Intellectual support for the individual will probably be provided by the ever-expanding computer networks which will greatly enhance our lives. Where our emotional and psychological support will come from is much less clear.

The new values now emerging will offer a challenge to each of us as individuals. Outsiders will no longer have the power to enforce their own moralities on us. We will need to experiment, as we started to do in the 1960s, with new forms of community. More directly, the onus is on us to choose carefully the lifestyle, or more likely the portfolio of lifestyles, to exactly match our unique needs at any point in time.

In summary, the first need here is for everyone to recognize the realities. As the first section in this summary explained, optimism rather than pessimism about the future should be demanded. The real need here is for grass-roots movements. Individuals will need to bypass the system and reassure each other that the future will be better. You, the reader, should tell all your friends, all your acquaintances and, especially, the politicians you know, that the future will be better; and you should demand that they agree!

Corporate Responses

When we look at how organizations might respond, the answer here is also simpler than you might expect. With more than five years of direct experience, backed by research among more than a thousand organizations, we have finally concluded that the main long-term problems for most organizations are not the result of incompetent management but are caused by a confusion of objectives. In these organizations, there should be at least two quite separate processes at work. There is, of course, the conventional corporate strategy process, optimizing performance in the shorter term, which we all know about. But there should also be a separate process of producing 'robust strategies' which underpin survival over the longer term; and it is these which relate to what you have read in this book.

These two quite distinct parts to the strategy process are best demonstrated by the table below, which we looked at in the first chapter, and which clearly establishes the significant differences between them:

	'Corporate' Strategy	'Robust' Strategy
Objectives	Optimizing Performance	Ensuring Survival
Characteristics	Short Term, Single Focus	Long Term, Divergent Coverage
Outcomes	Effective Commitment	Comprehensive Understanding
Beneficiaries	Individual Profiteers	Community Stakeholders

The two sets of strategies should have very different objectives. 'Corporate' (short-term) strategy is quintessentially about optimizing current performance, which requires that you find the single short-term solution which will deliver the optimal (internal) performance to which members of the organization can be persuaded to commit themselves most effectively.

'Robust' strategies, on the other hand, are above all about survival in the longer term, ensuring that all the potential threats are covered. They demand that multiple, and often divergent, objectives are met in order to exploit the potential emerging from changes in the external environment, and especially to guard against the whole range of threats which might endanger survival in the longer term, with the aim of understanding what these might be. This book has been intended to address these requirements.

There may be considerable tension between these two forms of strategy; and this is symptomatic of many of the stresses which are currently afflicting society.

Different Stakeholders

In the modern corporation, the individual 'shareholders' are no longer involved in managing the company themselves. Their investment in it may be as fleeting as a millisecond, as computers trade the shares on the electronic stock markets around the world. Understandably, therefore, their focus is only on the ephemeral performance of the share price, which may be more directly linked to stock-market rumours than to the financial performance of the firm. Individual senior managers of the organizations, the 'fat cat' beneficiaries, are interested in financial performance but again only in terms of the short-term results. Their own performance, and even more directly their income, is often tied to the share price. Both these sets of key 'stakeholders' demand that corporate strategy focuses on short-term financial performance.

The various communities hold a very different perspective. The employees of the organization have a very direct interest in the long-term survival of the organization, for on this depends their own survival. Even though many of them will, in time, change to jobs with other organizations, they still want the safety-net of a future in their current position. The other communities of stakeholders, from suppliers and customers through to the local communities which are dependent upon the organization, have similar requirements demanding long-term survival. With governments, such as the European Commission, increasingly recognizing the right of these communities to participate in the management of organizations, it is likely that such longer-term viewpoints may eventually prevail and robust strategies will finally come into their own.

Convergence of Strategies

I believe that the best answer to reconciling these differences in purpose and approach is simply to separate the two processes, the opposite of what is currently recommended. We call this process 'long-range marketing'. Fortunately, in practice the problem is usually not as acute as it might be expected to be. In our experience, despite the seeming contradictions, the two types of strategy typically converge on much the same approach, comprising a range of 'generic' long-term strategies, such as 'building relationships' with customers, which prove equally effective in both the shorter and the longer term.

Therefore, undertaking a separate identification of robust strategies does not necessitate a major revision of corporate strategy. In the relatively few cases where that is needed, the robust strategies clearly need to become the dominant part of the whole planning process. In general, there needs to be a new prioritization of existing strategy, with the emphasis subtly shifted to allow for the longer term in addition to the shorter one.

Simply by separating out the longer-term robust strategies from the shorter-term corporate strategy, organizations are better able to take account of the longer term, avoiding the problems which can arise from the short-termism generated by the pressures currently facing managements. The price they might have to pay, in terms of short-term steering, is usually small, whereas the long-term benefits may be great.

Government Responsibilities

Moving on to the topmost level, one key challenge to government is to manage the potential anarchy which could emerge from the fragmentation of society; for local government the challenge is to manage (and to resource) the all-important interface with each of us individually.

Beyond this, governments no longer need to accept artificial limitations on what may be achieved in the long term. In the short and medium term there will be bottlenecks which will limit certain courses of action, but they should beware of pessimistic statements that these transitional constraints will impose limits for ever. The responsibility on us, as individuals, is to recognize (and to punish) those politicians who refuse to recognize this obligation!

Governments, and organizations in general, will need to co-operate with each other as never before to build a shared future. They will have to face up to the realities underlying the potentially painful transition in just a few decades of many nations from the Third to the First World. This will need great statesmanship from the leaders of these Third World nations, but will pose even bigger challenges for the Western governments – especially that of the United States – which will have to face the ending of their hegemony and the start of their role as a minority.

Another challenge will be faced by national governments, which will need gracefully to cede some of their powers to governments at other levels. This is a major personal issue for the leaders involved, since politicians of every persuasion find any loss of power to be anathema. In theory, global problems should be resolved by the United Nations and its agencies. In practice, the historical limitations on these bodies will demand that other supranational bodies, especially those created by the new supranational groupings (including the EU and NAFTA), will need to address the most important issues.

In terms of the balance between nations, those in government will need rapidly to reorient themselves to the new realities. All of those determining the fate of the EU will need to recognize its responsibility as the most important new superpower. The United States will, on the other hand, need to recognize and rectify (if it can) its imminent descent into the abyss.

Within nations, the onus is on politicians to recognize, and address, the need for effective involvement of their electorates down to the level of each and every one of us, and then to implement the changes that are necessary to accomplish this. In practice, we will not just need to support the very few truly post-modern leaders such as Bill Clinton and Tony Blair, but we will have to wrest power from the old-style politicians by force. This will be, in a post-modern society based on one man one vote, through the ballot box, which will now be used to destroy the power of those parties which fail to comply with our wishes. It may, however, still involve some more direct action, through the growing number of single-issue groups and, as the last resort of a frustrated population, through civil unrest.

Global Expectations

There is on offer one set of tools which now may help governments to manage these massive changes more effectively. As was explained in the first chapter, the many individual 'predictions' made in this book were largely derived from the expectations which emerged from our research. Coming at these from a different direction, the main processes involved in these expectations can and should be managed to the advantage of individuals, organizations, governments and humanity as a whole.

Thus, the first concept, upon which much of the book was based, is that the future can now be predicted, as long as you properly understand the expectations of those many who will create it by their decisions. The corollary is that their expectations shape that future; and if you can influence those expectations you can influence the shape of the future.

The fundamental basis for this new capability is that the future is, in general, no longer constrained by the availability of scarce resources. The future is now determined by social decisions, and the outcome of these decisions can be observed in the expectations of those involved.

The concepts are encapsulated in the Rule of 3–2–1:

RULE OF 3–2–1

THREE assumptions:

1 *The future of humanity is, in general, no longer constrained by any significant shortage of resources.*

2 *Accordingly, that future is now being progressively determined by social decisions, taken not just by a few leaders but by millions of their citizens taking billions of small decisions as part of their daily lives.*

3 *The general, longer-term framework within which these specific, individual decisions is taken is largely provided by the individuals' expectations of what the future holds for them.*

TWO outcomes:

1 *If you can accurately measure these 'expectations' as to future developments, you can based on present conditions, equally accurately predict the most likely form of that future.*

2 *If you can shape these expectations, by whatever political or marketing processes available to you, you can equally shape that future and bend it away from the line it is currently following.*

ONE philosophy:

1 *All of us have the right, and the duty, positively to shape our own future and that of society as a whole.*

Governments must first know what are their electorate's existing expectations, the outcome of their previous conditioning by the environment in which they operate. Then they must either plan on the basis of the future which those expectations map, or they must shift them. If they successfully shift the expectations of a sufficient proportion to swing their overall decision, they will have determined the overall outcome and shaped the future. This is the greatest power they have over the creation of the future.

At the government level, the techniques are in many respects those already used by sophisticated marketing organizations except that the measurement of expectations may require special treatment. The use of modified forms of scenario forecasting, from which our own research results are derived, may often be the most suitable approach. The tension between existing expectations and desired outcomes may often be difficult to reconcile. Whatever the need, the techniques already exist – this book demonstrates that – but are not used!

Timescales

Even if the tensions are reconciled, all is not plain sailing. The first glitch is the sheer length of time it can take to persuade those involved. Those changes, for instance, which require a shift in culture, as will many of those described in this book, may take more than a decade to complete. This is much longer than most politicians and their electorates expect, and, indeed, is longer than their usual period in office! It is important, therefore, that they and especially we allow for this.

Legitimization

The persuasion will only work if it is believed, and the more believable it is the quicker the change will happen.

This legitimization will not, though, be the simple process of political 'persuasion' beloved of many politicians. This is a much more complex and, again, lengthier process than the quick fixes most politicians usually feel are sufficient. Underlying many of the current problems are the commitments by the establishment to an outdated set of values. Across the spectrum of society, from new individual lifestyles to new forms of family life and new forms of political life, our values are changing.

Perhaps the most important investment politicians will have to commit to is that of long-term education. This will be needed to persuade electorates, as well as themselves, to make the often painful changes which will be necessary to create the future. As we have seen, the shape of education will in future need to be very different from that traditionally deployed. The demand for lifelong education will become irresistible and will be recognized as the prime investment in the future of society and the route to empowerment of all of us.

A Better Future

As the final, final comment, I can offer no better advice than that you optimistically hope for a better future. More specifically, if you and the others around you genuinely expect a better future, it will come about!

REFERENCES

Material about the future of the world may be obtained from a wide variety of sources. Perhaps the most useful are the weekly news magazines such as *Newsweek* and *Time*; and the most useful of all is *The Economist*, which gives the widest and best coverage. Adding some specialist weeklies will extend the coverage: *Business Week* for management topics; *Scientific American* and *New Scientist* for science and technology. The reports of the leading think-tanks may also offer new ideas about the future.

Although the subjects referred to in this book are covered by many hundreds of books and journals, I have restricted myself here to listing just those publications which I have found most useful. The Futures Observatory web site includes details of the many more books and articles which have contributed to the work which led to the writing of this book. It can be found at:

http://oubs.open.ac.uk/future

Alternatively, you can talk to me, via email, at:

d.s.mercer@open.ac.uk

Adonis, Andrew and Geoff Mulgan, Back to Greece: The Scope for Direct Democracy, *Demos Quarterly*, No. 3, 1994

Amato, Ivan, The Sensual City, *New Scientist*, 15 October 1994

Armstrong, Karen, Fundamentalism, *Demos Quarterly*, No. 11, 1997
Atkinson, Dick, *The Common Sense of Community*, Demos, 1994
Avery, Dennis, 'The World's Rising Food Productivity' in *The State of Humanity*, ed Julian L. Simon, Blackwell, 1995
Badaracco Jr., Joseph L. and Allen P. Webb, Business Ethics: A View from the Trenches, *California Management Review*, Vol. 37 No. 2, Winter 1995
Beedham, Brian, Islam and the West, *The Economist*, 11 September 1993
Bergsten, C. Fred, The Rationale for a Rosy View, *The Economist*, 11 September 1993
Bezanson, Keith and Ruben Mendes, Alternative Funding: Looking Beyond the Nation-State, *Futures*, Vol. 27, No. 2, 1995
Bray, Francesco, Agriculture for Developing Nations, *Scientific American*, July 1994
Carmichael, Sheena, *Business Ethics: The New Bottom Line*, Demos 1995
Celente, Gerald, *Trends 2000: How to Prepare for and Profit from the Changes of the 21st Century*, Warner Books, 1997
Cetron, Marvin J. An American Renaissance in the Year 2000, *The Futurist*, March–April 1994
Cetron, Marvin J. with Owen Davies, The Future Face of Terrorism, *The Futurist*, November–December 1994
Coates, Joseph F., The Highly Probable Future: 83 Assumptions about the Year 2025, *The Futurist*, July–August 1994
Coates, Joseph F., Five Major Forces of Change, *The Futurist*, September–October 1996
Crook, Clive, The Future of Capitalism, *The Economist*, 11 September, 1993
Davidson, James Dale and William Rees-Mogg, *The Great Reckoning: How the World Will Change Before the Year 2000*, Pan Books, 1994
Drucker, Peter, *Post-Capitalist Society*, Butterworth-Heinemann, 1993
Eagar, Thomas, Bringing New Materials to Market, *Technology Review*, February/March 1995

Fukuyama, Francis, *The End of History and the Last Man*, Free Press, 1992

Galbraith, J. K., *The Affluent Society*, Hamish Hamilton, 1958

Gates, Henry Louis, Blood and Irony, *The Economist*, 11 September 1993

Handy, Charles, *The Age of Unreason*, Business Books, 1989

Harrison, Bennett, The Small Firms Myth, *California Management Review*, Spring 1994

Hallerstone, Simon, Apocalypse Now, or Maybe Later, *The Guardian*, 8 April 1995

Heilemann, John, Television, *The Economist*, 12 February 1994

Hobsbawm, Eric, *Age of Extremes*, Michael Joseph, 1994

Javetski, Bill and William Glasgall, Borderless Finance: Fuel for Growth, *Business Week*, March 1995

Kanter, Rosabeth Moss, Employability and Job Security in the 21st Century, *Demos Quarterly*, Issue 2, 1994

Kennedy, Paul, *Preparing for the Twenty First Century*, HarperCollins, 1993

Krugman, Paul, Does Third World Growth Hurt First World Prosperity, *Harvard Business Review*, July–August 1994

Krugman, Paul, *Peddling Prosperity*, Norton, 1994

Lansley, Stewart, *After the Gold Rush*, Century, 1994

Leadbeater, Charles and Geoff Mulgan, Lean Democracy and the Leadership Vacuum, *Demos Quarterly*, No. 3, 1994

Lean, Geoffrey, Too Small a World, *Independent on Sunday*, 28 August 1994

Lerner, Michael, *The Politics of Meaning: Restoring Hope and Possibility in an Age of Cynicism*, Addison-Wesley, 1996

Makridakis, Spyros, Mamagement in the 21st Century, *Long Range Planning*, Vol. 22, No. 2, 1989

Mautner, Michael, Engineering Earth's Climate from Space, *The Futurist*, March–April 1994

McRae, Hamish, *The World in 2020*, HarperCollins, 1994

McRae, Hamish, The Privilege of the Unemployed, *Independent on Sunday*, 26 February 1995

Mercer, David, The Foreseeable Future, *Management Decision*, Vol. 34 No. 3, 1996

Mercer, David, Marketing Practices in the 1990s, *Journal of Targeting, Measurement and Analysis for Marketing*, Vol. 5 No. 2, 1996

Merkle, Ralph C., It's a Small, Small, Small, Small World, *Technology Review*, February/March 1997

Milne, Janine, Break the Borders, *Computing*, 16 March 1995

Mulgan, Geoff, Networks for an Open Society, *Demos Quarterly*, No. 4, 1994

Mulgan, Geoff and Helen Wilkinson, Well-Being and Time, *Demos Quarterly*, No. 5, 1995

Naisbitt, John, *Global Paradox*, Nicholas Brealey, 1994

OECD, *The Knowledge-Based Economy*, OECD, 1996

Ogilvy, James, Future Studies and the Human Sciences: The Case for Normative Scenarios, *Futures Research Quarterly*, Vol. 8, No. 2, Summer 1992

Ohmae, Kenichi, Putting Global Logic First, *Harvard Business Review*, January–February 1995

O'Neill, Gerard K., *2081: A Hopeful View of the Human Future*, Jonathan Cape, 1981

O'Toole, Fintan, The Dredd of 2000 AD, *The Guardian*, 7 January 1995

Overholt, William H., *China: The Next Economic Superpower*, Weidenfeld & Nicolson, 1993

Paarlberg, Robert L., The Politics of Agricultural Resource Abuse, *Environment*, Vol. 3 No. 8, October 1994

Parker, John, Cities: Many Splendoured Things, *The Economist*, 29 July 1995

Pearce, Fred, Deserting Dogma, *Geographical*, January 1994

Pearson, Ian, *Technology Calendar: 1997–2045*, British Telecommunications, 1997

Pearson, Ian and Peter Cochrane, 200 Futures for 2020, *British Telecommunications Engineering*, Vol. 13, January 1995

Petersen, John, *The Road to 2015: Profiles of the Future*, Waite Group Press, 1994

Rockfellow, John D., Wild Cards: Preparing for 'The Big One', *The Futurist*, January–February 1994

Sagan, Carl, The Search for Extraterrestrial Life, *Scientific American*, October, 1994

Smith, Anthony, The Electronic Wars, *The Economist*, 11 September 1993

Stix, Gary, The Speed of Write, *Scientific American*, December 1994

Sudjic, Deyan, Megopolis Now, *The Guardian*, 24 June 1995

Toffler, Alvin, *Future Shock*, The Bodley Head, 1970

Toffler, Alvin, *The Third Wave*, William Collins, 1980

Toffler, Alvin, *Powershift*, Bantom Press, 1990

Toffler, Alvin and Heidi Toffler, *Creating a New Civilization*, Turner Publishing, 1994

Tough, Allen, *Crucial Questions About the Future*, Adamantine Press, 1995

Tutner, Richard, Hollywired, *The Wall Street Journal Europe*, March 1994

Wagar, W. Warren, *The Next Three Futures: Paradigms of Things to Come*, Adamantine Press, 1992

Ward, Mark, End of the Road for Brain Evolution, *New Scientist*, 25 January 1997

Wheat, Sue, Taming Tourism, *Geographical*, April 1994

Wilkinson, Helen, *No Turning Back: Generations and the Genderquake*, Demos, 1994

Wilkinson, Helen and Geoff Mulgan, *Freedom's Children*, Demos, 1995

Woodall, Pam, The Global Economy, *The Economist*, 1 October 1994

ACKNOWLEDGEMENTS

The justification for the views I put forward are not my own, but are those of the many hundreds of individuals who each gave so much time and energy to the work; and I am enormously grateful to each one of them. For the reader their participation is also the guarantee that this book has the authority necessary for you confidently to use the map of the future it describes as the basis for your own planning. It is the authority of those hundreds of contributors which should assure you that the views reported in the book are genuinely meaningful.

The first, and most important, acknowledgement, however, has to be to the anonymous participants in the research. An essential element was that confidentiality had to be guaranteed so that the individuals could speak their mind without fear that this would compromise their own position, or that of their own organization. Among these were the two hundred of our own MBA students, typically middle and senior managers from the larger organizations, who each spent between ten and twenty hours writing scenarios about the future of their own industry sectors; and in the process they contributed not just to our understanding of their industries but also helped us to refine the scenario-based techniques we were developing. Then there were the two hundred or so individuals, from the more than one hundred organizations, who participated directly in the twenty or so group discussions and scenario sessions which formed the core of the qualitative work. Each of these gave at least three hours of his or her time to a very intensive and demanding process. Finally, building on this group work, there were more than 300 individuals (from almost as many separate organizations) who completed a complex questionnaire covering more than 160 topics.

There also were the thousand or so other MBA students who, at various stages, participated in our additional questionnaire-based researches. Among those who also must remain anonymous were the 450 other outside participants, typically senior managers involved in corporate planning, who took part in our computer conferences. Some of them spent many hours over their contributions, which in total ran to several thousand hours spread over three years. These many hundreds of anonymous contributors, therefore, represent the ultimate authority for the expectations reported in the book. As such, I fully acknowledge the importance of their work and thank them for it. I hope their reward will be in a more stable, and more optimistic, future for their organizations.

In addition, a number of organizations have been more directly involved. Most obvious has been the internationally recognized Strategic Planning Society, which, under the aegis of Donald Alexander, jointly sponsored the earlier stages of the research and is now working with us on the Futures Observatory, which is continuing with the research. In addition, the highly regarded DEMOS think-tank led by Geoff Mulgan jointly sponsored the latter stages of the qualitative work. This work has been complemented by what we have now undertaken with the American Committee of the United Nations University (AC-UNU), led by Ted Gordon and Jerome Glenn. The Henley Centre for Forecasting, another leading think-tank, led by Bob Tyrrell, has given us extensive help, as has Shell, especially Graham Galer, who helped us develop the sophisticated techniques on which the research is based, and Roger Rainbow and Ged Davies, who contributed some of the key ideas.

In terms of the specific sections of the book, those on the Third World – as well as on national and international government – owe a great deal to members of the government of Ethiopia. Its president, Meles Zenawi, its prime minister, Tamrat Layne and especially its minister of defence, Seeye Abraha, spent many hours debating the concepts with me. The views of these individuals, who were among the most thoughtful but politically adventurous leaders in the world, have had a great influence on the shape of the book as a whole. It would take too much space to list all the other politicians and staff who, in any case, might be embarrassed to see their name in print in this context. They include members of governments in Japan (especially those involved with MITI) and the US, as well as those in Europe, and those in international bodies such as the EU (and especially members of its FAST programme, led by Ricardo Petrella, together with Michael Rogers, from Jacques Santer's office) and UNCTAD. To all of them I express my thanks for their contributions, which have, once again, shaped my reporting of the fields in which they operate. Similarly, within the United Kingdom I have been especially appreciative of the support provided by the staff at the DTI and the Foreign and Commonwealth Office, as well as the valuable input from ministers and staff at the Home Office, the Department of Health and the Cabinet Office, along with the executive team of the NHS (especially Philip Hadridge) together with various county and health authorities.

The advice and assistance of leading experts, from more than fifty other organizations, was especially appreciated during the earlier stages of the research. In this context, and in the specific area of the multinationals, I am once more especially appreciative of the input from Graham Galer from Shell, who led a very insightful debate on the future of such multinationals. Very valuable input, from opposing viewpoints, was also provided by the CEOs of both Toyota and General Motors. In this context, I am also grateful for the important contributions from Booker (Jonathan Taylor), British Airways (Brian Griffiths), BP (Peter Davies and Julian Chisholm), Fujitsu-ICL (Gill Ringland), Hanson (Stephen Park), ICI (Geoffrey Hobbs), and Unilever (David Stout). From the communications sector I would thank the CEOs of the main television networks in the

US, and especially Howard Stringer of CBS, for their individual contributions on the future of broadcasting, together with managers from IBM, ICL (Gill Ringland), NEC and DEC (David Skyrme), for those on the computer aspects of this subject. In the area of services, and especially retailing, I am particularly grateful to members of the staff of ASDA, especially Dominic Kelly, who spent many hours working with me; but I am also grateful for the influential contributions from Marks & Spencer (John Gardner), J. Sainsbury (Mark Venables), Waitrose (John Allan), Chef & Brewer (David Ankerson), as well as from Barclays (Chris Little and Kevin Wall) and Midland (Peter Stevens) banks and The Confederation of British Industries (Adair Turner and Sudhir Janankar). Not least, I am grateful to Ian Pearson, a futurist at BT (British Telecom, one of the very few organizations directly funding such activities), who contributed some of the most interesting and radical ideas, along with Graham Day at Leeds Metropolitan University and Bruce Lloyd at South Bank University, whose ideas also proved influential.

Finally, I would like to thank the representatives of the other major think-tanks for their own individual, and very influential, contributions. These include: the Policy Studies Institute, especially Jim Northcott and Nick Evans there; the Science Policy Research Unit, especially Peter Senke; PREST, especially Denis Loveridge; The Royal Institute of International Affairs, especially Oliver Sparrow; Futuribles (Hugues de Jouvenel) from France; The Copenhagen Institute of Future Studies (Martin Ågerup) from Denmark; and the Sekretariat Für Zukunftsforschung (Karlheinz Steinmüller) from Germany.

Where possible, I have expanded on the results of our own work by adding the most relevant contributions from a wide range of other commentators. As far as possible, these quotes are directly attributed to them. The list of publications referred to in this way is included in the References section.

In an academic context, I would like to thank all my colleagues, especially Rod Barratt, Harold Carter, Norman Fox, Margaret Greenwood, Jane Henry, Susun Mudambi, Alan Plath,

Tony Stapleton, Peter Stratfold, Andrew Thomson and Edith Thorne – as members of my course team – for their part in the stimulating debates which have taken place. I would also like to thank John Stopford, Robin Wensley and Sue Birley who gave me so many leads into the subject when I worked earlier with the London Business School. Needless to say, these colleagues, and many others, also contributed to the developments in education reported in the book.

I would like to thank the directors and senior managers at IBM who gave me such an insight into the subject, especially Derek Haslam, who gave me so much support over a number of years, and John Steele, who encouraged me to start my writing career.

As borough councillor and sub-committee chairman on behalf of a leading Residents Association, I would like to thank all my colleagues from the main political parties, as well as from the Association itself, who gave me the practical evidence to support many of the more adventurous ideas about politics reported in the book, as well as the academics in the Tavistock Institute with whom I initially worked on these ideas.

Finally, I come to those who contributed so much hard work supporting the various stages of the project: to Martin Bartle, who managed the work with DEMOS; and to Joanne Tyler, who arranged much of the work with the Strategic Planning Society. Then there are Mo Vernon and Val Page, course managers with the Open University, who helped me set up and run the Millennium Project – and did so much to make it a success – and the staff in the Open University library who helped me to find the hundreds of books and, quite literally, thousands of articles which fed into the project and thence into this book. There are also Giles Clark and Jonathan Hunt in the book publishing department of the OU, who gave such helpful advice on each of the many revisions the book went through. Almost last, but definitely not least, my thanks to my secretaries Gloria Rippin and Karen McCafferty, who have provided the support needed to keep me going over the more than seven years which the main research has taken. A major part of any success is owed to the indefatigable efforts of my agent, Sonia Land. I cannot thank her enough for her literary advice and business guidance, as well as finding me the ideal publisher.

INDEX